GENDERCIDE

New Feminist Perspectives

GENDERCIDE

The Implications of Sex Selection

MARY ANNE WARREN
San Francisco State University

ROWMAN & ALLANHELD
PUBLISHERS

ROWMAN & ALLANHELD

Published in the United States of America in 1985
by Rowman & Allanheld, Publishers
(a division of Littlefield, Adams & Company)
81 Adams Drive, Totowa, New Jersey 07512

Library of Congress Cataloging-in-Publication Data

Warren, Mary Anne.
  Gendercide : the implications of sex selection

  (New feminist perspectives)
  Includes index.
  1. Eugenics.   2. Sex preselection.   3. Sex of
  children, Parental preferences for.   4. Feminism.
I. Title   II. Series
HQ751.W37   1985   305.3   85–14452
ISBN 0–8476–7330–8
ISBN 0–8476–7334–0 (pbk.)

85 86 87 / 0 9 8 7 6 5 4 3 2 1
Printed in the United States of America

To utopian feminists,
who dream of alternative futures.

Contents

Acknowledgments

I am grateful to Mary Vetterling-Braggin, who invited me to submit a proposal and offered encouragement during the early stages of the project. It is impossible to mention individually all the others to whom thanks are owed. I am deeply indebted to the many feminist scholars and writers who have cast light upon the hidden past and highly uncertain future of women, in both western and nonwestern cultures. I have tried to acknowledge these debts at appropriate points in the text, but I hope I may be forgiven if faulty memory has led to a failure to give credit in every instance in which it is due. I have learned much from discussions with students and colleagues at Sonoma and San Francisco State universities, Saint Olaf College, and the University of Western Australia. My friends in the Pacific Division of the Society for Women in Philosophy and the Feminist Discussion Group in Perth have provided invaluable support for the pursuit of philosophical issues that have not generally been recognized as legitimate by mainstream philosophers.

I would also like to thank all those at Rowman & Allanheld who participated in the various stages of the book's production. They have coped patiently with my frequent migrations between California and Australia and with the resulting confusions and delays.

Last, but far from least, my heartfelt thanks to my spouse, Michael Scriven. He has been unfailingly supportive, even when I have used him as a sounding board for ideas that many men might have found threatening. His example has lent strength to my conviction that moral philosophy must be a broad, interdisciplinary subject, drawing upon all available sources of knowledge in its approach to the ethical issues that confront our species.

Introduction

The idea that millions of women and female children have been killed or allowed to die because of their sex would probably strike most people as false or even paranoid. The further suggestion that sexually discriminatory deaths continue to occur in many parts of the world—including our own society—is apt to be met with ridicule or indignation. Such widespread ignorance is not surprising, given the continuing lack of awareness of the wider phenomenon of sexism. Even many supposedly well-educated people find it easy to dismiss sexism as a relatively trivial problem. Many admit that there was once a great deal of unjust discrimination against women, while insisting that things have now changed to the point that it is primarily men who need to be defended against sexism.

The fact is that sexism is often a matter of life and death for its female victims. Throughout recorded history, in virtually every patriarchal society, innumerable female human beings have been killed, starved, or otherwise abused as a result of the cultural devaluation of the female sex. The neglect or abuse of females has often been severe enough to produce societies in which men greatly outnumber women. I use the term "gendercide" to refer to those wrongful forms of sexual discrimination which reduce the relative number of females or males, whether through direct killing or in more indirect ways. (Actions which are not morally wrong are not gendercidal, even if they happen to affect sex ratios. For instance, if a program to improve the nutrition of pregnant women resulted in relatively more male infants being born, this would not be a form of anti-female gendercide.)

Not all gendercide is anti-female gendercide. Millions of men have also died in part because of their sex, e.g., in wars for which only men are conscripted. Wars which decimate the male population of a nation or community may be regarded as a *de facto* form of gendercide. But there are some crucial differences between anti-male and anti-female gendercide. Men are not sent to fight wars because of female-dominated institutions or matriarchal ideologies which depict men as inferior because of their sex. On the contrary, the male monopoly of the conduct of war is arguably a consequence of male domination in patriarchal societies. Moreover, it tends to contribute to the maintenance of that domination.

The Amazons, warrior women depicted in ancient Greek art and legend, were said to have killed or abandoned their male offspring. But we do not know whether these Amazons ever existed. If they did, the Greeks—who took credit for defeating them and eliminating their form of life—are hardly a reliable source of information about their customs. Most authenticated instances of anti-male gendercide, such as the killing of male captives in war (sparing women and children for a life of slavery) have occurred in male-dominated societies.

The victims of anti-male gendercide are often members of a conquered or oppressed racial, religious or economic group, and their deaths are as much a form of genocide as of gendercide. One possibly mythical instance related in the Bible is the killing of male infants in Israel, said to have been ordered by the Roman governor, who feared the birth of a rebel leader. A more recent example is the lynching and discriminatory execution of black men in the United States. In such cases too, anti-male gendercide is not based upon a cultural de-valuation of *all* males.

The evidence for the prevalence of anti-female gendercide is clear and unambiguous. It can be denied only through ignorance, prej-udice, or both. Such ignorance or prejudice leaves us ill prepared to face the crisis presented by the emergence of new methods for selecting the sex of children before birth or before conception. It is possible that within the next generation the relative number of women on the planet will be vastly reduced as a result of these new methods of sex selection. If this should happen, the consequences for women's social, political and economic status might be devastating. For history makes clear that although women (and men) sometimes profit from a shortage of their own sex, the "rarity value" of women often leads to imprisonment rather than power.

This book is written from a feminist perspective. To say this is not to admit a bias, but to declare a belief in certain facts that I hold to be objectively demonstrable, and certain values I regard as morally essential. The basic empirical belief that unites all feminists is that women of many societies have been and still are severely oppressed because of their sex. The basic moral belief is that this oppression is wrong and must be ended. I have no exclusive loyalty to any one branch of the feminist movement, such as radical feminism, spiritual feminism, socialist feminism, or lesbian feminism. Each of these elements of the women's movement has made important contributions to feminist theory, and I hope that I have benefited from the insights of all of them.

Feminists do not seek to replace male domination with female domination, but to achieve equal rights and equal justice for persons of both sexes. Sexual inequality is only one of the massive problems facing humanity. Yet it is intimately linked to all the others, from the proliferation of nuclear weapons, international imperialism, economic injustice, and racial and religious conflict, to overpopulation and the destruction of nonhuman species and natural environments.

What stance should feminists take toward the development and use of new methods of sex selection? Should sex selection be condemned as a threat to feminist goals? Should it—as many have argued—be welcomed as a means of slowing down the too-rapid growth of the human population? Or are all such predictions about the long-term social consequences of sex selection premature, and therefore to be discounted? Do prospective parents have a right to have access to present and future methods of sex selection, regardless of the risk that the widespread use of these methods may have undesirable consequences? Might the new methods of sex selection be used in ways which will *not* be harmful to women, but which will increase their freedom?

This book addresses these questions and others that arise along the way.

Chapter 1 begins with an overview of the currently available and possibly soon-to-be-available methods of sex selection. Next, it explores the concepts of sexual equality and gendercide, and briefly explains some of the reasons why feminists are concerned about the development of new methods of sex selection. Finally, it poses the moral dilemma created by the emergence of new techniques which will enable parents to select the sex of children before their birth.

Chapter 2 presents some basic facts about female infanticide and other forms of anti-female gendercide, both in the past and in the present. The persistence of such practices, and of the misogynist ideologies which serve to rationalize them, demonstrates the urgency of the concern that sex selection may be used in gendercidal ways.

Chapter 3 explores some of the more extreme options posed by the emergence of sex selection and other new reproductive technologies. Might women be eliminated altogether, once the development of ectogenesis (i.e., the extrauterine gestation of fetuses) makes the female sex unnecessary for human reproduction? At the opposite extreme, might women cease to have sons once cloning, parthenogenesis or ovular merging make *men* unnecessary for reproduction? Might *both* sexes be eliminated, in favor of some intermediate form? Far more probable than any of these radical options is the formation of self-reproducing communities consisting entirely of women or entirely of men. But none of these options is a realistic one for the vast majority of people. The arguments for and against the moral and legal toleration of sex selection must be considered primarily in the context of sexually mixed families and communities.

Chapter 4 addresses those objections to sex selection which do not depend upon any particular predictions about its social consequences. Some of these are equally applicable to either preconceptive sex selection or sex-selective abortion. For instance, is sex selection morally objectionable because it is contrary to nature? Is it "playing God"? Is it inherently sexist to want to preselect the sex of a child? Other nonconsequentialist objections apply only to sex-selective abortion. Is abortion itself morally wrong? Is sex selection an acceptable reason for abortion? And is late (i.e., second- or third-trimester) abortion for the purpose of sex selection any less morally objectionable than sex-selective infanticide?

Chapter 5 addresses what may be the most alarming of the consequentialist objections to sex selection: that by increasing the relative number of males it will lead to a more violent world. There is some evidence—though no conclusive proof—that human males are by nature more apt to learn aggressive and violent forms of behavior than are human females. If this hypothesis is correct, will more males necessarily mean more violence?

Chapter 6 deals with the other consequentialist objections to sex selection. Will women suffer a loss of rights and freedoms because of relative increases in the number of males? Will girls be psychologi-cally harmed by the use of sex selection to produce male firstborns? Will sex selection thwart women's efforts to gain control over their own lives by increasing their dependence upon the male-dominated medical profession? Will an increasingly male population result in even more destructive exploitation of the natural world? Will it increase class differentiation, or result in a higher proportion of women born into poverty? Will the "excess" men who cannot find wives be unhappy? And might sex selection and other new reproduc-tive technologies lead to more dangerous forms of human genetic engineering?

Chapter 7 deals with the possible benefits of sex selection, such as avoiding sex-linked diseases, reducing birth rates, increasing the happiness of parents, and avoiding the birth of children who will be abused or neglected because of their sex. Chapter 8 presents the case for individual freedom of choice in the use of the new methods of sex selection.

I will argue that the objections to sex selection are insufficient to show that it is inherently immoral to preselect the sex of a child. The nonconsequentialist objections to most forms of sex selection are unsound. The consequentialist objections are based upon predictions about the long-term consequences of sex selection which may turn out to be false; and the possible benefits cannot be ignored. The one form of prenatal sex selection which is inherently objectionable is late-term sex-selective abortion. Yet even this is something for which we ought not to blame those individual women who may be desperate enough to choose or consent to it.

This does not mean that feminist concerns about *all* of the new methods of sex selection—not just sex-selective late abortion—are ill-founded. There is a real danger that these methods will be used in ways which prove harmful to women and to men as well. The use of these new methods may either help or hinder the cause of sexual equality. In all probability, it will do both, in different ways. The actions taken by the women's movement may have a decisive influence upon the social impact of the new methods of sex selection. The Conclusion focuses upon some of the things which must be done to minimize the negative effects of sex selection.

1

The Prospect of Sex Selection

New Methods of Preselecting Sex

The folklore of most cultures is filled with recipes for preconceptive and prenatal sex selection, most of them designed to increase the odds of male births. Letty Pogrebin lists just a few of these recipes for the production of males:

> The couple should have intercourse in dry weather, on a night with a full moon, after a good nut harvest, and/or when there is a north wind.
>
> The man should wear boots to bed, get drunk, tie a string around his right testicle, cut off his left testicle, take an ax to bed, hang his pants on the right bedpost, and/or bite his wife's right ear.
>
> The woman should lie on her right side during intercourse, eat red meat or sour foods, let a small boy step on her hands or sit in her lap on her wedding day, sleep with a small boy on her wedding eve, wear male clothing to bed on her wedding night, and/or pinch her husband's right testicle before intercourse.[1]

Of course, we know now that such folk methods are ineffective. But modern science promises to provide much more reliable methods of selecting the sex of children. Prenatal sex selection is already a reality. Already there is a reliable means of selecting sex after conception but prior to birth, i.e., through prenatal sex diagnosis and abortion of fetuses of the "wrong" sex. Early sex-selective abortion is not yet widely available, but it may soon become more so. Moreover, it may soon be possible for prospective parents to reliably preselect the sex of a child prior to its conception.

The sex of the human individual is determined at conception, when an ovum is fertilized by either an androgenic (male-producing) or a gynogenic (female-producing) spermatozoon. If the ovum is fertilized by an androgenic sperm it develops into a male fetus, which has a

Y-chromosome in one of its 23 chromosome pairs; if it is fertilized by a gynogenic sperm it gives rise to a female fetus, which has two X-chromosomes. For the past decade, it has been possible to diagnose fetal sex prior to birth through amniocentesis. In this procedure, a needle is inserted into the uterus through the women's abdominal wall, and a sample of amniotic fluid is withdrawn. This fluid contains fetal cells, which may be cultured and their chromosome structure examined to determine the fetus's sex.

Amniocentesis cannot safely be performed in the first trimester of pregnancy, since the amniotic sack is too small and the risk of damaging the fetus too great. Furthermore, the test results often take several weeks to obtain, as tissue cultures must be grown. Thus, if sex-selective abortion is desired, it must be performed well into the second trimester. Few prospective parents, at least in the industrial-ized nations, are willing to abort a pregnancy which is this far along simply because of the fetus's sex. There is widespread (though not universal) agreement among genetic counselors and medical ethicists that prenatal sex diagnosis should not be offered solely for the purpose of sex selection, except when there is a serious danger that a male child would suffer from a severe sex-linked ailment such as hemo-philia.[2] There have been reports in the United States of sex-selective abortions following amniocentesis, but the numbers are apparently very small.[3] In India, however, sex-selective abortion is becoming quite common. In Punjab physicians have erected billboards adver-tising amniocentesis for sex diagnosis, prompting protests by feminists.[4] For, as everyone admits, virtually all of the fetuses aborted because of their sex are female.

Fetal sex can also be detected by ultrasonic scanning, although this is usually not possible until the third trimester. Skilled ultrasonogra-phers can detect the shape of the fetus's genitals in a fairly high proportion of cases (about 66 to 74 percent).[5] In many parts of the United States and Europe, ultrasonic scanning is routinely used to check the fetus's position and gestational age, and the parents are often told its sex. Because third-trimester abortions are rarely performed in these countries, except for much more compelling reasons, this method has apparently not yet been used for the purpose of sex selection. However, two researchers at the University of California at San Francisco have developed a method of detecting fetal sex by ultrasonic scanning during the second trimester, which they claim is 100 percent accurate.[6] If this method is as accurate as reported, and if it comes into wider use, ultrasonography may sometimes be used to facilitate sex selection.

Other methods of detecting fetal sex which are currently being developed or tested promise to permit even earlier results. Of these, the most promising is chorion biopsy. In this procedure, a sample is taken through the cervix of the chorionic villi, the small rootlike projections on the chorion, or outer portion of the placenta. The biopsy may be performed several weeks before the end of the first trimester, and highly accurate results may be obtained within a few days.[7] Chorion biopsy has been in use for a decade in China, though few figures are available as to how often it has been used for sex selection.[8] It is now being tested in the U.S. If this or other methods of early fetal sex detection become more widely available, sex-selective abortion may become much more common throughout the world.

A reliable method of preconceptive sex selection would, however, be much more attrative to most prospective parents. Whatever one's view of the moral status of abortion, it is inherently unpleasant. Early abortion is many times safer than childbirth, but still carries some slight risk. Abortions performed past the midpoint of pregnancy may be almost as dangerous as childbirth. As will be argued in Chapter 4, late abortions are morally more problematic than early abortions, and tend to be emotionally traumatic for all concerned. The risks of properly performed abortion are not sufficient to induce most women to complete pregnancies which occur at a time when they do not want to have a child at all; but they become more significant when a child is wanted and the fear is only that an already conceived fetus may not be of the preferred sex.

At present, no reliable method of preconceptive sex selection is available for use on human beings. But this situation may change. There are many theories about ways in which the sex of children can be influenced prior to conception. Most are still highly speculative, but one or more may eventually give rise to a reliable method of preconceptive sex selection.

One theory is that the timing of intercourse and conception can alter the odds of producing a male or female child. Because androgenic spermatozoa tend to be more numerous, and because they have been thought to be shorter-lived and faster-moving than gynogenic spermatozoa, it has been hypothesized that intercourse close to the time of ovulation is more apt to produce male conceptions, while intercourse several days prior to ovulation is more apt to produce females. In a popular book on sex selection published in 1970, David Rorvik and Landrum Shettles recommended that couples wanting boys should have intercourse close to the time of ovulation, while those wanting girls should have intercourse several days prior to that time.[9] However Elizabeth Whelan, in a competing book, argued that this schedule actually *lowers* the probability of getting a child of the desired sex.[10] Studies have been done which tend to support each of these contradictory theories. To confuse matters even further, some researchers have claimed that the appropriate schedules are reversed in the case of artificial insemination. Guerrero has reported that natural inseminations early and late in the fertile period produce more males than those nearer to the time of ovulation, while the opposite effect occurs with conceptions resulting from artificial insemination.[11] It is entirely possible that the timing of conception does sometimes affect the sex of the offspring; but at present the evidence is so inconsistent that no method of sex selection involving timing appears particularly promising.[12]

A second theory is that acidic environments are more favorable to gynosperm, while alkaline environments favor androsperm. Rorvik and Shettles recommend the use of acid douches to increase the odds of conceiving a girl, and alkaline douches to increase the odds of a boy.[13] On the same theory, couples wishing to conceive a boy have been advised to use deep penetration, since the secretions of the cervix are thought to be less acidic than those of the vagina. These methods, like those involving the timing of conception, have not been proven to be reliable. There appears to be no valid evidence that pH values in the female reproductive tract have any effect upon sex ratios.[14]

A third theory is that fetal sex can be influenced by the mother's diet in the weeks prior to conception. Stolkowski and Choukroun advise that a woman who wants to conceive a boy should eat foods high in sodium and potassium; for a girl, she should eat foods high in calcium and magnesium.[15] The assumption is that a woman's internal mineral balance may affect the consistency of her cervical mucus, or some other environmental condition within her reproductive tract, making it more hospitable to sperm of one or the other sort. Several other researchers have recommended particular diets for the production of boys or girls.[16] So far, however, there has been no experimental confirmation of such claims, and most fertility researchers regard the odds of selecting sex through diet as close to nill.

A fourth theory is that higher sperm counts are conducive to the conception of males. On this theory, men who want sons should be healthy and well nourished, wear loose clothing around their testicles, and avoid ejaculating for several days prior to the attempted conception. Repeated intercourse on the same day apparently also increases the sperm count within the female tract. (The Talmud advises men who want sons to have intercourse with their wives twice in succession.)[17] Experimental confirmation of this theory is also scant, although it enjoys wide popular acceptance.

None of these "home remedies" for the production of boys or girls have been shown to be effective. But medical technology may yet develop highly effective methods of preconceptive sex selection. There has been speculation about the eventual development of a sex-selection pill which might, for instance, alter the ratio of androgenic and gynogenic sperm produced by the man, or induce the woman's immune system to selectively attack and destroy sperm of one or the other sort. A method has been patented to isolate the separate antibodies to androgenic and gynogenic sperm which are thought to be produced by the ovum, but it has not thus far been marketed.[18] Barrier methods, such as a diaphram which would allow only one type of sperm to pass through, have also been suggested. However, little or no progress has been made towards the development of such a method.

There is, however, one preconceptive method of sex selection which is already technically possible, i.e., in vitro fertilization and the implantation of an embryo of the desired sex. Thanks largely to the work of Robert Edwards and Patrick Steptoe, in vitro fertilization has become a well-established procedure, available in England, Australia, the United States, and many other parts of the world. It involves the removal of one or more ova through a small incision in the woman's abdomen; these are then externally fertilized with sperm from her husband or another donor and introduced into her uterus. ("In vitro" means "in glass," and refers to the contrast with natural fertilization, which occurs in vivo, i.e., inside the woman's body.) At present, the majority of the ova fertilized in vitro and introduced into the uterus fail to implant. Even when several zygotes are introduced, pregnancy often fails to occur and the procedure must be repeated. However, the success rate is said to be gradually improving, and further improvements are likely.

In vitro fertilization is currently used to aid women who cannot conceive in the usual way, due to blocked fallopian tubes or other physiological abnormalities. But it could also be used for sex selection. In the earliest stages of its development, when it consists of only a few cells, the conceptus may be divided, apparently without much danger of inducing later abnormality. Thus, a portion could be removed and examined to determine its sex, and the remainder used to establish a pregnancy. It is unlikely, however, that in vitro fertilization will be used simply for sex selection so long as the resources available for it remain as limited as they still are. Under present circumstances, overcoming infertility will continue to take precedence over selecting sex.

But the most promising methods of preconceptive sex selection are probably those involving the separation of androgenic and gynogenic spermatozoa, and artificial insemination using sperm of predominantly one type. A number of methods of sperm separation have been developed and tested with an eye to use in the commercial breeding of animals. These include centrifugation,[19] electrophoresis,[20] sedimentation,[21] and albumin isolation.[22] The results of the use of these methods in animal breeding have not been impressive. However, the albumin isolation method, developed by Ronald Ericsson, is currently available for human use in about two dozen clinics around the world.[23] This method exploits the allegedly greater motility of androsperm. Sperm are placed above a denser medium, such as bovine serum albumin. More of the androsperm are thought to penetrate this medium, and artificial insemination is performed with the supposedly androgenic-enriched sperm. Ericsson claims a success rate of about 77 percent for the conception of boys by this technique. However, this claim has not been independently verified, and is doubted by many other researchers.

Methods of separating gynogenic sperm are also being developed. The Philadelphia Fertility Clinic is testing a method in which semen is placed in a glass column filled with tiny beads of Sephadex gel. It is thought that the androgenic sperm are more likely to adhere to the beads, and thus that more of the gynogenic sperm reach the bottom of the tube.[24] The number of conceptions which have been produced by this method is still very small, and more work is needed before its efficacy can be established.

It would be rash to predict that dramatic breakthroughs in techniques of sperm separation are immanent. Such predictions have been made for decades and thus far the results have been disappointing. Nevertheless, it is entirely possible that the continuing research in this area will eventually pay off, perhaps within the next few years.

Still another means of sex selection might arise from the future discovery of ways of inducing parthenogenesis. In parthenogenetic reproduction, the ovum begins dividing without fertilization by a spermatozoon. Many invertebrate species (e.g., aphids, flatworms and certain arthropods) routinely reproduce asexually through parthenogenesis, as do several species of fish and lizards. It is possible that parthenogenesis may occasionally occur spontaneously in humans, though no proof of this speculation is available. Because the human ovum has only an X-chromosome, the resulting embryo could only be female—a fact which may help to explain why so little research has been done on human parthenogenesis. However, parthenogenetic reproduction has been induced in amphibians through a number of different methods, such as pricking the eggs with a needle or treating them with antibiotics or ions. A strain of turkeys has been developed in which many unfertilized eggs begin to develop, and a few grow to maturity.[25]

It may also eventually prove possible to induce two human ova to fuse and begin cell division, thereby producing a daughter with two mothers and no father. Pierre Soupart has reported success in inducing ovular merging in mice.[26] If this report is correct, then the prospect of success with our own species may be better than has generally been assumed. Methods of inducing parthenogenesis or ovular merging would be of great interest to single or lesbian women, who might wish to conceive without heterosexual intercourse or the use of donated sperm.

The Preference for Sons

Once safe and effective methods of preconceptive sex selection exist, they are apt to be widely used. If these methods prove expensive, or if they require sophisticated skills and equipment, they may initially be confined to the more wealthy industrialized nations. But an inexpensive and relatively simple method would be apt to spread rapidly to most parts of the world. For throughout the world prospective parents tend to have distinct preferences about the sex of their future children.[27] In most societies, they tend to prefer sons over daughters.[28] Son-preference involves some or all of the following desires: (1) that if there is to be just one child in the family, that one be male; (2) that if there are to be several children, there be more males than females, or only males; and (3) that the firstborn child be male. Women as well as men generally tend to prefer sons, although the son-preference of men is usually more pronounced.

Son-preference is not a culturally uniform phenomenon. Nancy Williamson has published two extensive cross-cultural surveys of the studies which have been done on son-preference. It is clear that son-preference tends to be much stronger in the Third World nations of Asia, Africa, and Central and South America than in the industrialized nations, where the most common preference is for one child of each sex.[29] There are even a few societies in which daughter-preference is predominant.[30] Yet even in the United States and Western Europe there is still a pronounced tendency to prefer male children, particularly among men, and particularly with respect to the firstborn child. Although the status of women has improved somewhat over the past decades, recent studies of sex preferences indicate that the preference for sons remains more or less unchanged.[31]

Son-preference is a direct result of patriarchy. By "patriarchy" I mean the type of social system which ensures male control of women and (other) resources. This control is maintained through many devices, including (1) patriliny and patrilocality;[32] (2) the denial of women's rights to sexual and reproductive autonomy, ownership and inheritance of property, education, freedom of movement, gainful economic employment, and political and religious participation; and (3) overt violence against women or girls, or the threat thereof.

There are many forms and degrees of patriarchy, and some have doubted the value of the concept as a cross-cultural descriptive category. It is all too easy to project western assumptions onto nonwestern cultures. Feminist anthropologists have demonstrated that much of the ethnographic literature, e.g., about gathering-and-hunting societies, has been distorted by the androcentric biases of male anthropologists.[33] Yet it is also true that almost every known culture exhibits some degree of male domination, and that this domination is usually maintained through at least some of the means just named.

In general, the lower women's social, legal and economic status, the stronger will be the tendency of both women and men to prefer sons. Some daughter-preferring societies are matrilineal and/or matrilocal, and most of them seem to be sexually egalitarian to an unusual degree.[34] Matriliny and matrilocality tend to contribute to sexual equality, and thus to increase the perceived value of women. They do not, however, lead to matriarchy, i.e., the economic, social and political domination of women. There are no known matriarchal societies.

Son-preference also tends to diminish in response to economic development.[35] Industrialization often coincides with the reduction of the dependence of elderly persons on support from children (usually sons), as governments increasingly assume such social-security functions. Technological changes in the modes of production and the spread of education to women as well as men often increase the economic opportunities available to women, while undermining traditional assumptions about women's roles. Yet, as Williamson notes, son-preference remains strong even in some societies, such as Korea and Taiwan, which are experiencing rapid economic development.[36]

The degree to which parents in a given society tend to value sons more than daughters provides a measure of the extent to which women's rights are denied and suppressed. Of course, a person's *value* (to other persons) is one thing, and her *rights* are another. Some slaves may be more highly valued than others, without necessarily being allowed any more extensive rights. Nevertheless, children's potential value to their parents is largely determined by the rights and opportunities which they will later enjoy. Patriarchy ensures that in most cases sons will be likely to provide their parents (or be *thought* to provide them) with a greater return on their investment. Sons, unlike daughters, can be expected to carry on the family name, to augment the family's economic resources, and to defend its interests in the public arena.

In a patriarchal system, a daughter is usually only a temporary member of her family of origin, since she will leave it as soon as she is old enough to marry. Her children, like her labor, will belong to her husband's family, not that of her father or mother. Even if her husband is required to pay a bride-price to her parents, this is not usually enough to cover the cost of her upbringing. If, on the other hand, her parents are required to provide a dowry, the cost of raising daughters is apt to be perceived as far greater than any resulting benefits. In India, the demand for large dowries still constitutes a powerful disincentive to the raising of daughters. Although people obviously do not have children solely for the sake of economic gain, it would be naive to doubt that such considerations are often of great importance.

Given that most societies throughout the world and throughout recorded history have been patriarchal, it is little wonder that while the birth of a son has usually been greeted with joy, the birth of a daughter has more often been an occasion for ill-concealed disappointment. As Pogrebin points out,

> Son preference has been so universal for so long that jokes about it are commonplace: "I'll send it back if it's a girl," a man tells his wife as she is wheeled into the delivery room. A nurse shows a man his newborn baby and the man says, "That's OK. A girl was my second choice."[37]

Joking about such disappointment is one way to make it seen less important. But son-preference has a darker side, in the murder or abuse of daughters and women who fail to bear sons. Nearly everyone knows about King Henry VIII of England, who annulled his marriage to Catharine of Aragon when, after twenty years and numerous miscarriages, she still had not given him a male heir. His second wife, Anne Boleyn, fared even worse: when she gave him "only" a daughter—the future Queen Elizabeth—he had her beheaded.

Son-preference is as old as patriarchy, and unlikely to vanish any sooner. Sex selection is nothing new either. Female infanticide, and the selective neglect, abuse, or abandonment of female children have been common occurrences in most historical eras, and have by no means entirely disappeared. In many contemporary societies, the lower survival rates for female children demonstrate the persistence of such postnatal forms of sex selection. In India, women have a higher death rate throughout their lives, and many more girls die in infancy.[38] The higher death rates of female infants cannot be explained simply as the result of inadequacies in the diet or medical care of infants of both sexes, since these conditions would be more apt to produce a higher mortality rate among male infants, who tend to be more vulnerable to disease than females.

It is not only economic systems based on private property which generate such abuses. In the People's Republic of China, socialism has somewhat lessened the inequality of women without entirely eliminating patriarchal customs and attitudes. There, the government has found it necessary to institute educational programs to persuade parents to raise their female children, rather than killing or abandoning them.[39] Faced with an enormous and still expanding population, China has instituted increasingly severe penalties for families with more than one child. This, together with the persistence of son-preference, has made many people reluctant to raise daughters.

Sex Selection and Sex Ratios

The persistence of son-preference means that if an effective and easily affordable means of sex selection were to be made generally available, the most probable immediate consequence would be an increase in sex ratios. Sex ratios are commonly represented as the number of males per 100 females. Thus, higher ratios represent higher relative numbers of males. Although this usage may reflect the assumption that the more males the better, it seems best to retain it here. (After all, "higher" does not always mean "better"!)

This increase in sex ratios will probably be most marked in those societies where patriarchy is strongest, and where religious beliefs are least apt to inhibit the use of the new methods. China, Taiwan, North and South Korea, India, Pakistan, Bangladesh, and many of the nations of northern and central Africa and the Middle East would probably experience the largest sex-ratio increases. In these nations, particularly in the rural areas, sex-preferences among both men and women are often as high as four to one in favor of males, and sometimes they are even higher.[40]

There are many reasons why such figures cannot be assumed to constitute an accurate indication of the sex ratios that would result from the introduction of affordable methods of sex selection. All of the many sex-preference surveys have found a sizable proportion of persons who are unwilling to state their sex-preferences (if any). Many of these persons may nevertheless *have* preferences, and be willing to implement them. There is no easy way to test the truthfulness of stated preferences. Furthermore, it is difficult to predict what proportion of those with stated preference would be willing to use sex-selection methods in order to implement those preferences, or which *sorts* of methods they would be willing to use. A 1977 survey of students at five northern California colleges found that 44.6 percent would like to make use of sex-selection methods.[41] This is a somewhat higher proportion than that found in most previous studies,[42] and probably reflects an increasing acceptance of the idea of sex selection. It seems likely that the level of acceptance will continue to increase in most of the world, as knowledge of the new reproductive technologies spreads. However, it is also possible that there will be a backlash against these new technologies, and that the willingness of prospective parents to select their children's sex will decrease.

In spite of these uncertainties, it is clear that the potential exists for enormous sex-ratio changes in many nations. It is sometimes argued that any sex-ratio imbalances resulting from sex selection would be slight and temporary. Westoff and Rindfuss, for instance, claim that,

> If effective sex control technologies were rapidly and widely adopted in the United States ... the temporary effect would be a surplus of male births in the first couple of years. This would be followed by a wave of female births to achieve balance, and the oscillations would eventually damp out. Ultimately, under conditions of sex predetermination, the sex ratio would be similar to the existing natural sex ratio at birth of 105.[43]

This argument assumes that parents will automatically act so as to produce a fairly even sex ratio. But why should we assume this? The argument usually given is that a shortage of women will increase the perceived value of female children, making parents more eager to have girls. In India, however, there are just 930 females per 1000 males, and this imbalance has not led to an increased desire to have girls. On the contrary, the imbalance has increased over the past several decades. Son-preference is largely a result of the social and economic disadvantages suffered by females. High sex ratios will not automatically eliminate those disadvantages, and thus will not necessarily undermine that preference. As Helen Holmes points out: "To change a low proportion of females in any population the underlying low regard for females and economic advantage of males must be changed."[44]

In Europe and the United States, son-preference is relatively weak, but still prevalent. Studies done between the 1930s and the 1970s show that the most common preference is for one child of each sex, though most often the son is wanted first. If only one child is wanted, a male tends to be preferred; and if more than two children are wanted there is a strong tendency to prefer a predominance of males. Williamson reports that, "Preferences for a single or first girl or for a predominance of girls are rare," although about a third of persons interviewed claim to have no sex preferences.[45]

Although son-preference tends to be stronger in the Third World nations, there are some notable exceptions. Indonesia, Thailand, and the Philippines show comparatively low levels of son-preference. Although son-preference—especially on the part of males—remains fairly strong in the predominantly Roman Catholic nations of Central and South America, two factors exist which are apt to moderate sex-ratio increases due to the introduction of new methods of sex selection in these nations. First, as Williamson's survey shows, "women of Spanish background appear to favor girls as much as they do boys."[46] Evidence from the Caribbean area, for instance, suggests that although men tend to prefer sons, women are just as apt to prefer daughters. A study of Peruvian women revealed preferred sex ratios ranging from 93 to 116, showing only a slight tendency towards son-preference.[47] The second factor is the influence of the Roman Catholic church.

It may seem ironic that an institution which is committed to male-supremacist doctrines should oppose the implementation of son-preference. Yet this will almost certainly be the case. Because it remains opposed to abortion for any purpose, the Church will inevitably oppose sex-selective abortion. It is also likely to oppose most preconceptive methods of sex selection, such as those involving in vitro fertilization or sperm separation and artificial insemination. For these methods require that sperm be obtained through mastur-bation, which the Church considers morally objectionable, even when the sperm is later used to produce conception. Many church officials also oppose in vitro fertilization because it often entails the death of embryos which are fertilized but not introduced into the uterus, or those which are introduced but which fail to develop. Methods which do not interfere with the normal method of concep-tion (e.g., those involving diet, the use of acid or alkaline douches, or the timing of conception) would not be seen as objectionable for *these* reasons. But it is likely that sex selection itself will be viewed by the Church as a usurpation of divine prerogatives.

This argument is essentially the one which Pope John Paul II has used against the practice of most forms of contraception. According to John Paul,

> At the origin of every human person there is a creative act of God ... man and woman are not the arbiters, are not the masters of this same capacity ... When, therefore, through contraception, married couples remove from the excercise of their conjugal sexuality its potential procreative capacity, they claim a power which belongs solely to God.[48]

If contraception is viewed as a usurpation of divine rights, it seems likely that attempts to select the sex of a child will be viewed in the same light. (The validity of this type of argument will be considered in Chapter 4.) Many Catholic couples will probably disregard the Church's views on sex selection, just as many now defy its prohibition of contraception and abortion. But just as the Church's influence has contributed to the maintenance of high birth rates in many predominantly Catholic nations, so it will probably suffice to moderate the effect of sex selection upon sex ratios.

None of the other major world religions is as firmly opposed to abortion, in vitro fertilization and other reproductive technologies as is Catholicism. Thus, if new methods of sex selection become generally available, we may expect the most extreme sex-ratio changes to occur in the most severely patriarchal of the non-Catholic nations. However, *some* increase in sex ratios is likely to occur in virtually every nation in which effective methods of sex selection become widely available.

The Threat to Sexual Equality

Feminists embrace the goal of sexual equality, i.e., of equal—and adequate—rights, freedoms and opportunities for women and men. Or rather *most* feminists embrace this goal. Those who reject it generally do so on the grounds that the concept of moral equality, as it has developed in the context of western patriarchal society, does not and cannot do full justice to the special needs and interests of women.[49] My view, in contrast, is that the ideal of moral equality— properly understood and freed from its original context in western law and philosophy, where it applied only to white males—must be a fundamental postulate of any satisfactory moral system.

Sexual equality is a special instance of the wider ideal of the moral equality of all persons. Three points need to be made in order to clarify the ideal of sexual equality. First, sexual equality does not imply that women and men should always be treated in exactly the same way, regardless of their differing needs and interests. To the extent that women and men have different needs, sexual equality requires that these different needs be met, so far as this is possible and just.

Second, the ideal of sexual equality does not imply that there are no psychological or behavioral differences between women and men, or that there ideally ought to be no such differences. Some feminists endorse the ideal of "psychological androgyny" (or "gynandry"),[50] hoping for the eventual elimination of all general psychological differences between the sexes, and all stereotyped expectations about how women and men ought to differ in their behavior. Other feminists reject this ideal because they believe that women and men are *naturally* different, in more than just the obvious physical ways, and that these differences need to be respected.[51]

Third, sexual equality is not a substitute for racial, economic, and international justice. Feminists are fully aware that sexist discrimination is closely associated with unjust discrimination on the basis of race, religion, age, sexual preference, international imperialism, and other forms of oppression. The elimination of sexual inequality requires the simultaneous elimination of these other injustices.

What is crucial to the ideal of sexual equality is the claim that *equal consideration should be given to the legitimate needs and interests of persons of both sexes*. In patriarchal societies, men's needs and interests generally take precedence over those of women. To achieve the goal of sexual equality, women need to be more equally represented in all important social, political, and economic institutions. Moreover, these institutions themselves must be radically altered. Hierarchical systems which force individuals to compete for a few positions of power and affluence must be transformed by more egalitarian modes of organization. Few feminists believe that much can be accomplished simply by integrating a few women into existing male-dominated institutions—although this is often a necessary first step. After several thousand years of patriarchy, much of the world has begun to make erratic but significant progress towards the goal of sexual equality. But sex selection might threaten or even contribute to the reversal of this progress.

It is important to realize that unequal sex ratios are not intrinsically unjust, and do not necessarily violate the ideal of sexual equality. It is not obvious that either sex has a collective moral right to constitute at least half of the total human population. The only plausible way to demonstrate the existence of such a right would be to show that if one sex is in the numerical minority, members of that sex will necessarily be unfairly disadvantaged. We know that this is false in the case of males, since at present men are a numerical minority (approximately 49 percent) in the industrialized nations, but they are not visibly disadvantaged as a result of this fact. The question of whether it is also false in the case of women will be considered in Chapter 6.

There may, indeed, be some circumstances in which higher sex ratios would prove beneficial to women. A relative scarcity of women might increase their value in both the marriage and labor markets, thereby enabling a strong feminist movement succesfully to demand more equitable laws and social arrangements. But where women do not already possess substantial legal and political rights, or where the women's movement is weak, disorganized or nonexistent, higher sex ratios would be less likely to work to their advantage. Under these conditions, a shortage of women might lead instead to a tightening of patriarchal restrictions, as many men come to believe that maintaining control over the behavior and reproductive lives of women is more important than ever.

The preference for male firstborns is of as much concern to feminists as the preference for all-male or predominantly male families. There is some evidence that firstborn children tend to enjoy significant advantages in personal, social, and intellectual development.[52] The claim that birth order is significant in itself has recently been challenged. But even if birth order is inherently insignificant, there is still room for concern that girls will suffer from the knowledge that they were *selected* to be born only after the preferred sons.

Another concern is that, if sex selection becomes available primarily to wealthy families, it may lead to an even greater concentration of wealth and power in male hands and an increasing proportion of women born into poverty.[53] Feminists have also pointed out that the physiological risks of sex selection (e.g., the risks of second-trimester abortion), will be suffered primarily by women. Moreover, the widespread adoption of the practice of selecting children's sex will be apt to increase women's dependence upon the male-dominated medical profession.

These are just a few of the reasons why many feminists fear that sex selection will prove inimical to the goal of sexual equality. But it is well to admit at the onset of our investigation that we cannot know in advance what the social consequences of sex selection will be. We cannot even be certain that it will have a significant effect upon sex ratios. If reliable and inexpensive methods of preconceptive sex selection are not developed and made widely available, if there is doubt about the safety of the new methods, or if the most effective methods continue to require invasive interventions into the reproductive process, then sex selection may never become much more widespread than it is at present. It is possible (though very unlikely) that by the time reliable methods of sex selection become widely available son-preference will have become much less common. It is also possible that by then the son-preference of some will be counterbalanced by the daughter-preference of others, e.g., feminists and women who choose to raise a child without a male partner.

The Genocide Analogy

Many of the moral issues raised by the prospect of sex selection may usefully be posed through an analogy between the concept of *genocide* and what I call *gendercide*. The *Oxford American Dictionary* defines genocide as "the deliberate extermination of a race of people."[54] By analogy, gendercide would be the deliberate extermination of persons of a particular sex (or gender). Other terms, such as "gynocide" and "femicide," have been used to refer to the wrongful killing of girls and women. But "gendercide" is a sex-neutral term, in that the victims may be either male or female. There is a need for such a sex-neutral term, since sexually discriminatory killing is just as wrong when the victims happen to be male. The term also calls attention to the fact that gender roles have often had lethal consequences, and that these are in important respects analogous to the lethal consequences of racial, religious, and class prejudice.

The concept of genocide, as it is commonly understood, does not apply only to those actions which result in the complete extermination of a race of people—such as very nearly occurred to the natives of Tasmania at the hands of the nineteenth-century English colonists.[55] Sometimes it is appropriate to speak of certain actions as genocidal atrocities, even though many members of the victimized race or culture survive. The Holocaust, in which millions of Jewish people within the Nazi-occupied territories of Europe were systematically murdered, is the paradigm case of genocide, and the case which shocked the international community into a récognition of genocide as a crime against humanity.[56]

Furthermore, not all instances of genocide involve direct or deliberate killing. Deaths or cultural disintegration deliberately or negligently brought about through starvation, disease or neglect may also be genocidal. Indeed, some acts of genocide do not involve any deaths at all, but rather consist in the wrongful denial of the right to reproduce. The involuntary and discriminatory sterilization of black and Native American women, for instance, must be regarded as an instance of genocide, even though it has not typically caused the deaths of any members of those groups.

To call such actions genocidal is to condemn them as morally wrong. Genocide is a crime not just against certain individuals or groups but against humanity as a whole. When a people or a culture is destroyed, humanity is deprived of the contributions which that group might otherwise have made. However, not all actions which result in a relative decrease or the prevention of an increase in the number of persons of a particular race or culture are instances of genocide. If that were the case, it would follow that every time a woman decided not to have a baby, or another baby, she would be contributing to a process of genocide. This is absurd, since no woman is morally obligated to have as many children as she possible can; indeed, given the threat of overpopulation, she may be morally obligated not to.

In the light of these considerations, I would suggest that an action, law or policy should be regarded as genocidal if (1) it results in an absolute or relative reduction in the number of persons of a particular racial or cultural group; and (2) the means whereby this result is brought about are morally objectionable for independent reasons— e.g., because they violate certain individuals' rights to life, liberty, or security against wrongful assault. On this criterion, involuntary sterilization, contraception or abortion, imposed upon persons because of their race or culture, is genocidal; but the voluntary and uncoerced use of such measures is not.[57]

The uncoerced use of birth control methods is not immoral, and hence not genocidal, because individual persons and couples have a moral right to make such decisions regarding their own reproductive lives. If some persons were to make use of birth control methods for morally dubious reasons (e.g., a malicious desire to deprive their parents of the joys of having grandchildren), they would still be acting within their moral rights, and thus might not be guilty of genocide. Indeed, even in the unlikely event that *all* the members of some racial group were to voluntarily opt for permanent childlessness, they might still not have committed genocide, but only racial suicide. If they were driven to this choice by desperation born of oppression, then perhaps their oppressors could be accused of genocide; but they themselves would not necessarily have done anything immoral.

If "genocide" means wrongfully killing or otherwise reducing the relative number of persons of a particular race, then "gendercide" means the same thing, except that "sex" is substituted for "race." Like genocide, gendercide need not involve outright murder, although the paradigm examples of it do. Like genocide, gendercide involves actions which are morally objectionable for reasons apart from the mere fact that they may cause an alteration in the numerical ratios between certain groups.

Gendercide is no less a moral atrocity than genocide. Those who object to the genocide/gendercide analogy on the grounds that it seems to belittle the genocidal crimes which have occurred in the past, and which continue to occur in our own time, greatly underestimate the severity of gendercidal crimes in both the past and the present. As the next chapter will show, the number of human beings who have been killed because of their sex is probably just as great as the number who have been killed because of their race or religion. When gendercide involves murder, it is wrong for that reason, but for additional reasons as well. These include the implied insult and threat to all members of the victimized sex, which is always present when gendercidal practices are justified through sexist ideology, and the loss to humanity of the contributions which might otherwise have been made by the victims.

However, not all actions which happen to reduce the relative number of persons of one sex or the other are gendercidal. Improved medical care has differentially increased the average life span of women in the industrialized nations, thereby reducing the relative number of men alive at the present time; but this is not an instance of gendercide, since it is not morally objectionable to provide people with improved medical care. If men's currently shorter average life spans in most of the industrialized nations can be shown to be the result of systematic injustices against them, then there might be grounds for speaking of gendercide. But at present this appears not to be the case. It seems probable that men now tend to die younger primarily because of their greater natural susceptibility to many potentially lethal diseases.[58] This greater susceptibility to disease is evident in males even before birth. Many more males than females are conceived, but male fetuses are more apt to be miscarried, perhaps because they are more apt to have some genetic defect. It is also possible that the high-stress life style of many contemporary men contributes to their earlier deaths, and that this high stress level is related to the contingencies of the male gender role. If so, then this may be a subtle instance of gendercide, inflicted upon males by a male-dominated culture.

The central questions facing us may now be stated as follows: Is sex selection, especially when it results in imbalanced sex ratios, a form of gendercide? Is it gendercidal in some instances and not others? Is there a moral difference between preconceptive sex selection and selective abortion, such that the one is morally acceptable but the other must be viewed as a form of gendercide? If so, why is this difference so significant? And does it make a crucial moral difference whether sex-selective abortions are performed early or late in the pregnancy?

Once we have answered these questions, we must face the additional question of what feminists should do to minimize the gendercidal consequences of sex selection. If we decide that sex selection is morally objectionable, should we seek to place legal restrictions or prohibitions upon its development or use? Some actions which are morally questionable nevertheless ought to be legally permitted, e.g., because the results of prohibition are apt to be on the whole worse than the results of toleration. And conversely, some actions which are not harmful in isolated individual instances may nevertheless need to be prohibited because the frequent occurrence of these actions would have undesirable consequences. Thus, the morality of sex selection and the desirability of prohibiting it are two distinct questions.

The Risks and Benefits of Freedom

There are valid reasons for concern about the social consequences of sex selection. But it would be a mistake to ignore the possible benefits which sex selection may also provide. Used *against* women, sex selection could become part of a process by which progress towards sexual equality is permanently halted. At the logical extreme, women might be eliminated entirely: fetuses could be gestated in artificial wombs, and ova produced in mechanically supported isolated ovaries. Used *by* women, on the other hand, it could be part of the process by which, in Shulamith Firestone's words, women seize control of the means of reproduction (their own bodies and their own parenting), thereby making it possible to replace the patriarchal family with more egalitarian social structures.[59]

It is, after all, a significant part of women's oppression that they have had to bear and raise sons, who will become part of the patriarchal power structure. Women will soon have at least the theoretical option of refusing to do this, without avoiding parenthood altogether, or abandoning their male children, as the Amazons were said to do. Some women may choose to form self-reproducing unisex communities: the ultimate lesbian separatist adventure. Perhaps relatively few women will be attracted by such an option. But those who believe that women's abilities flower best in a community of women should value the opportunity to test this hypothesis within transgenerational all-female families or communities.

It would be foolhardy to predict that the overall results of sex selection will be beneficial to women, or to the goal of sexual equality. Yet legal right and the technical means to choose the sex of one's children will still be of some value, to women as well as men. It will, at the very least, make possible new modes of rebellion against male domination.

How do we weigh the risks and benefits of freedom against those of prohibition? This question has had to be faced in connection with every potentially revolutionary technological innovation. New reproductive technologies—from effective contraception and medically safe abortion to artificial insemination and in vitro fertilization—have always seemed to the conservative mind to be particularly threatening. Such technologies can alter the form and the very constitution of the family unit—the mainstay of patriarchal society as we know it. Even though these new technologies have often subjected women to new medical risks (while enabling us to avoid some of the old risks associated with sex and reproduction), it is still plausible to hold that at present their net effect upon women's lives is generally beneficial. But sex selection might prove to be the exception to this rule. How can we know, and how should we react in the light of the radical uncertainty of its long-term consequences?

No philosopher has ever devised a fully objective and reliable method for determining the appropriate moral or legal response to an innovation whose social consequences are so difficult to predict. Utilitarians advise us to determine the long-term consequences of each option and endorse that which will be likely to produce the maximum amount of happiness or satisfaction for the greatest number, with the minimum of suffering and frustration. But when we cannot know the long-term consequences of our actions, or even which particular set of consequences is the most probable, utilitarianism is of little use. On a right-based theory, we ought to first determine what our moral rights are, and then defend those rights even if we cannot know that doing so will maximize the long-term benefits. But how can feminists decide whether or not we ought to claim the right to select the sex of our children when we do not even know whether the recognition of that right would tend, on the whole, to increase or decrease the opportunities for further progress towards sexual equality? Kantians tell us to determine the morality of an action by asking whether we could rationally will that everyone should act in that way. But how can we decide whether it is rational to will that sex selection be used by all prospective parents (or all those who wish to preselect their children's sex), when we cannot know what the results of such universal use would be?

None of these theories offers us much guidance with respect to so complex an issue as sex selection. Rather than seeking to determine the moral status of sex selection through the application of one of these formal moral theories, I will adopt the more pragmatic procedure of first exploring the context within which sex selection must be viewed, and then considering each of the moral objections to sex selection in the light of this overall context.[60] The history and continued occurrence of anti-female gendercide is part of this context, as are the extreme options for the human future which could conceivably come about as a result of sex selection and other new reproductive technologies.

Notes

1. Letty Cottin Pogrebin, *Growing Up Free: Raising Your Child in the 80's* (New York: Bantam Books, 1981), 82.

2. *Screening and Counseling for Genetic Conditions*, a report by the President's Commission for the Study of Ethical Problems in Medicine and Biomedical and Behavioral Research (Washington D.C.: U.S. Government Printing Office, 1983), 57–58.

3. John C. Fletcher estimates that the total number of amniocenteses done for sex selection in the approximately 125 American prenatal diagnostic centers between 1970 and 1983 does not exceed 50. ["Ethics and Public Policy: Should Sex Choice Be Discouraged?", in *Sex Selection of Children*, edited by Neil G. Bennett (New York: Academic Press, 1983), 226.] The actual figure may be somewhat higher, since deception is possible: prospective parents can sometimes obtain sex diagnosis through amniocentesis by falsely claiming a family history of hemophilia or some other sex-linked disease.

4. See Arun Chacko, "Too Many Daughters? India's Drastic Cure," *World Paper*, November 1982, 8–9; and Viola Roggencamp, "Abortion of a Special Kind: Male Sex Selection in India," in *Test Tube Women: What Future for Motherhood?*, edited by Rita Arditti, Renate Duelli Klein, and Shelly Minden (London: Routledge and Kegan Paul, 1984), 266–77.

5. See Lachlan C. de Crespigny and Hugh P. Robinson, "Determination of Fetal Sex with Ultrasound," *Medical Journal of Australia* 2 (July 25, 1981): 98–100; and E. M. Weldner, "Accuracy of Fetal Sex Determination by Ultrasound," *Acta Obstetricia et Gynecologia Scandinavica* 60 (1981): 333–34.

6. John D. Stephens and Sanford Sherman, "Determination of Fetal Sex by Ultrasound," *The New England Journal of Medicine* 309 (1983): 984.

7. J. R. Gosden et. al., "Direct Vision Chorion Biopsy and Chromosome-Specific DNA Probes for Fetal Sex Determination in First Trimester Prenatal Diagnosis," *The Lancet* 2 (December 25, 1982) 1416–19.

8. Tientung Hospital, "Fetal Sex Prediction by Sex Chromatin of Chorionic Villi Cells During Early Pregnancy," *Chinese Medical Journal* 1 (1975): 117. This article reports that, of the 30 abortions requested after fetal sex diagnosis, 29 were of female fetuses.

9. David Rorvik and Landrum B. Shettles, *Your Baby's Sex: Now You Can Choose* (Toronto: Dodd, Mead and Company, 1970).

10. Elizabeth Whelan, *Boy or Girl?* (Indianapolis: Bobbs-Merrill Company, 1977).

11. R. Guerrero, "Sex Ratio: A Statistical Association with the Type and Time of Insemination in the Menstrual Cycle," *International Journal of Fertility* (1970), 221–225.

12. For a summary of research on the influence of the timing of conception, see William H. James, "Timing of Fertilization and the Sex Ratio of Offspring," in *Sex Selection of Children*, 73–95.

13. Rorvik and Shettles, op. cit.

14. James, 89.

15. J. Stolkowski and J. Choukroun, "Preconception Selection of Sex in Man," *Israel Journal of Medical Science* 17 (1981), 1061–67. A report on this work also appeared in *People Magazine*, 1982.

16. See, for instance, Sally Langendoen and William Proctor, *The Preconception Gender Diet* (New York: Evans and Company, 1982); and J. Lorrain and R. Gagnon, "Selection Preconceptionelle du Sexe," *L'Union Medicale du Canada* 104 (1975), 800–803.

17. Marcia Guttentag and Paul F. Secord, *Too Many Women? The Sex Ratio Question* (Beverly Hills: Sage Publications, 1983), 100.

18. C. Campbell, "The Manchild Pill." *Psychology Today*, August 1976. 86; cited by M. Ruth Nentwig, "Technical Aspects of Sex Preselection," in *The Custom-Made Child?: Woman-Centered Perspectives*, edited by Helen B. Holmes, Betty B. Hoskins, and Michael Gross (Clifton, New Jersey: The Humana Press, 1981), 335.

19. Centrifugation, or rapid rotation, tends to cause the slightly heavier gynosperm to separate from the lighter androsperm; however, the procedure may also reduce the viability of the sperm.

20. Electrophoresis involves subjecting sperm to an electric current, which tends to cause the gynosperm to cluster around the anode and the androsperm around the cathode.

21. In the sedimentation or agglutination technique, the heavier gynosperm are allowed to sink to the bottom of a column.

22. See Ronald J. Ericsson and Robert H. Glass, "Functional Differences Between Sperm Bearing the X- or Y-Chromosome," in *Prospects for Sexing Mammalian Sperm*, edited by Rupert P. Amaun and George E. Seidel Jr. (Boulder: Colorado Associated University Press, 1982), 201–11.

23. *Time*, August 27, 1984, 67.

24. Stephen L. Corson, Frances R. Batzer, and Sheldon Schlaff, "Preconceptual Female Gender Selection," *Fertility and Sterility* 40:3 (1983), 384–85.

25. See Jeremy Cherfas and John Gribbon, *The Redundant Male: Is Sex Irrelevant in the Modern World?* (London: The Bodley Head, 1984), 40.

26. Sally Miller Gearhart, "The Future—if There Is One—Is Female," in *Reweaving the Web of Life: Feminism and Nonviolence*, edited by Pam McAllister (Philadelphia: New Society Publishers, 1982), 282. Gearhart provides a provocative discussion of the possible social benefits of reproduction through parthenogenesis or ovular merging.

27. See, for instance, L. C. Coombs, "Preference for Sex of Children Among U.S. Couples," *Family Planning Perspectives* 9 (1977), 259; C. H. Coombs, L. C. Coombs, and G. H. McClelland, "Preference Scales for Number and Sex of Children," *Population Studies* 29 (1975), 273–98; Simon Dinitz, Russell R. Dynes, and Alfred C. Clarke, "Preferences for Male or Female Children: Traditional or Affectional?" *Marriage and Family Living* 16 (May 1954), 128–30; and Gerald E. Markle, "Sex Ratio at Birth: Values, Variance, and Some Determinants," *Demography* 11 (1974), 131–42.

28. Nancy E. Williamson, "Boys or Girls? Parents' Preferences and Sex Control," *Population Bulletin* 33 (1: January 1978); and *Sons or Daughters: A Cross-Cultural Survey of Parental Preferences* (Beverly Hills: Sage Publications, 1976.)

29. Williamson, *Sons or Daughters*; and J. E. Clare and C. V. Kiser, in *Social and Psychological Factors Affecting Fertility*, edited by P. K. Whelpton and C. V. Kiser (New York: Milbank Memorial Fund, 1951), 621–73.

30. Williamson, *Sons or Daughters*, 103–16.

31. See Mary E. Pharis and Martin Manosevitz, "Sexual Stereotyping of Infants:

Implications for Social Work Practice," *Social Work Research and Abstracts* 20 (1984), 7–12.

32.   Patriliny is the inheritance of family names, family membership, and property through the male line. Patrilocality is the practice according to which the husband determines where the married couple will live, e.g., with the former's family or clan. Most highly patriarchal societies are both patrilineal and patrilocal, although there are some exceptions, and neither patriliny nor patrilocality is invariably a sign of a highly patriarchal culture.

33.   See, for instance, Diane Bell, *Daughters of the Dreaming* (Melbourne: George Allen & Unwin, 1983). Bell presents evidence that central Australian aboriginal women, long assumed to have been excluded from the politics and religion of their cultures, have important religious and cultural traditions of their own, and enjoy a much greater degree of personal autonomy and political influence than has previously been suspected. The patriarchal mindset of previous (mostly white male) observers, and their assumption that male informants could tell them all they needed to learn about aboriginal cultures, rendered women's autonomous activities nearly invisible.

34.   Williamson, *Sons or Daughters*, 103–16. The daughter-preferring societies which Williamson mentions are the Mundugumor of Papua New Guinea, the Tiwi of North Australia, the Garo of Assam, the Iscobakebu of Peru, and the Tolowa Indians of Northwestern California.

35.   Bennett, 8.

36.   Williamson, "Boys or Girls?" 29.

37.   Pogrebin, 84.

38.   See Barbara D. Miller, *The Endangered Sex: Neglect of Children in Northern Rural India* (Ithaca: Cornell University Press, 1981), 68–82.

39.   In 1983, Prime Minister Zhao Ziyang of China made a speech condemning the practice of female infanticide—a fairly clear indication of its persistence or resurgence in China. (Bennett, 9.)

40.   Nancy E. Williamson, *Sons or Daughters*, 74–88.

41.   Shirley Foster Hartley and Linda M. Pietraczyk. "Preselecting the Sex of Offspring: Technologies, Attitudes, and Implications," *Social Biology* 25:2 (1978), 232–46.

42.   See, for instance, Charles F. Westoff and Ronald R. Rindfuss, "Sex Preselection in the United States: Some Implications," *Science* 184 (May 10, 1974), 636; and Gerald E. Markle and C.B. Nam, "Sex Predetermination: Its Impact on Fertility," *Social Biology* 18 (1971), 73–82.

43.   Westoff and Rindfuss, 633.

44.   Helen B. Holmes, "Sex Preselection: Eugenics for Everyone?" forthcoming in *Biomedical Ethics Reviews, 1985*, edited by James Humber and Robert Almeder (Clifton, New Jersey: Humana Press).

45.   Williamson, *Sons or Daughters?*, 33.

46.   Williamson, "Boys or Girls?", 13.

47.   Williamson, *Sons or Daughters?*, 69.

48.   Pope John Paul II, speech to participants in a study seminar on responsible parenthood, September 17, 1983. Cited by George Cook, Director, Catholics United for the Faith, in *The Australian*, March 1984.

49.   See Elizabeth H. Wolgast, *Equality and the Rights of Women* (Ithaca: Cornell University Press, 1980).

50.   Some feminists prefer the term "gynandry" to "androgyny," because it puts the female term first, and because it is less apt to be confused with "androgeny," which means a condition caused by excessive exposure to male hormones.

51.   For further discussion of the feminist ideal of psychological androgyny, see

"Femininity," "Masculinity," and "Androgyny," edited by Mary Vetterling-Braggin (Totowa, New Jersey: Littlefield, Adams and Company, 1982).

52.   Some research has suggested that firstborn children tend to be more intelligent, more self-confident, and more successful, and that women with older brothers are particularly severely disadvantaged in these respects. See W. D. Altus, "Birth Order and Its Sequelae," *Science* 151 (1966), 44. Other researchers have concluded that the alleged advantages of firstborn status are largely an artifact of improper control procedures; see John Tierny, "The Myth of the Firstborn," *Science 83* 10:10 (December 1983), 16. Chapter 6 provides a more extensive discussion of research on the psychological effects of birth order.

53.   See Roberta Steinbacher, "Futuristic Implications of Sex Preselection," in *The Custom-Made Child*, 196.

54.   *Oxford American Dictionary* (New York: Avon, 1980), 272.

55.   The last full-blooded Tasmanian aborigine died in 1876, and for nearly a century the history books taught that the race was entirely extinct. In fact, there are several thousand descendants of the original Tasmanians, who still regard themselves as belonging to that race and culture.

56.   The term "genocide" was coined by Raphael Lemkin, who successfully lobbied for an international convention on genocide, passed by the United Nations in 1946. See Leo Kuper, *Genocide: Its Political Use in the Twentieth Century* (New Haven: Yale University Press, 1981), 23–24.

57.   Of course, if abortion were itself generally immoral then there might be a basis for regarding even voluntary abortion as a form of genocide. In Chapter 4, I argue against the view that abortion is inherently immoral.

58.   For a partial list of the diseases which affect males more often than females, see Ashley Montagu, *The Natural Superiority of Women* (New York: Collier, 1974), 87–88.

59.   Shulamith Firestone, *The Dialectic of Sex: The Case for Feminist Revolution* (New York: Bantam, 1971), 10–11.

60.   It is not my purpose here to develop an abstract theory of the nature of morality. However, I believe that to be satisfactory such a theory must incorporate both consequentialist and nonconsequentialist elements. The concept of a moral right, for instance, is one which I believe to be essential, and which probably cannot be adequately explicated in strictly consequentialist terms.

2

Gendercidal Precedents

It is impossible to evaluate the implications of the new methods of sex selection without an awareness of the prevalence of gendercide throughout recorded history. An understanding of some of the forms which gendercide has taken in the past may facilitate a recognition of the forms in which it survives today, and may persist into the future. The material in this chapter will not be unfamiliar to feminist scholars, and may safely be skipped by those who are already aware of the many forms of anti-female gendercide. It is included in part as a response to those who may regard the very concept of gendercide as a symptom of paranoia.

The writings of Andrea Dworkin,[1] Mary Daly,[2] and other feminist investigators have made us more aware of such gendercidal atrocities as the European witch hunts of the Renaissance and Reformation periods, *suttee*, and the genital mutilation of women. Somewhat less attention has been paid to the practice of female infanticide and the selective abandonment or neglect of female children. Yet these are even more common forms of gendercide, both historically and in the contemporary world. They are also more directly relevant to our own investigation, since they are essentially nontechnological forms of sex selection, and thus may illustrate some of the dangers posed by the newer forms of sex selection.

Female Infanticide

Few people today are aware of how common the practice of infanticide has been in most cultures and in most periods of history. Even fewer realize that where infanticide is practiced in a patriarchal context, female infants are nearly always the most frequent victims.

Patriarchy tends to create conditions in which the raising of daughters is a less profitable investment than the raising of sons. Consequently, daughters are often regarded as an unbearable expense, and often they are either eliminated shortly after birth or allowed to die of neglect.

Sex selection is not the only reason for infanticide, even in patriarchal cultures. In many tribal societies, an infant would be killed if the mother was unmarried, if she died in childbirth, or if it was sickly, abnormal, or born in a manner considered to be unpropitious.[3] Among nomadic gathering-and-hunting groups, an infant might be killed because the mother already had one small child and could not manage with two.[4] Sometimes, as in the precolonial Hawaiian Islands, the usual practice was to kill all children after the third or fourth.[5] In precolonial Tahiti, the members of the lowest classes were reportedly required to kill nearly all of their children, regardless of sex.[6] In India and throughout much of the ancient Middle East and Europe, infanticide was practiced as a sacrificial rite. Firstborn sons were sometimes killed as a religious ritual, or because they were regarded as a danger to the father;[7] this was the fate which nearly befell Oedipus. Sometimes infants were killed for medicinal or magical purposes, or buried in the foundations of a new building in the belief that this would give it strength.[8]

But though there have been many reasons for infanticide, sex selection has long been one of the most common. The Hellenistic author Posidippus remarked that, "Even a poor man will bring up a son, but even a rich man will expose a daughter."[9] That statement is to some degree applicable not only to most of the pre-Christian cultures of the Middle East and Mediterranean region, but to the majority of all patriarchal cultures which have not enforced a general prohibition of infanticide.

The practice of infanticide has not been confined to patriarchal societies. Some societies which are quite sexually egalitarian, such as the !Kung people of southern Africa, have practiced infanticide.[10] Conversely, some very strict patriarchal traditions, such as Judaism, Islam, and Christianity, have prohibited infanticide—though never with complete success. It is probable that our human and protohuman ancestors practiced infanticide long before the advent of patrilineal systems of descent and other social institutions constitutive of patriarchy in its more modern forms.

Darwin believed that infanticide is a specifically human phenomenon, and a perversion of the natural reproductive instinct which
could not possibly predate what he called civilization, and what we
might call culture.[11] But we know now that animals of many
species—including those primate species which are biologically
closest to our own—sometimes destroy their own young or those of
conspecifics. This occurs under natural conditions and not only in the
unnatural conditions of captivity.[12] Given that our earliest human
ancestors were gatherer-hunters, and that most of the gathering-
and-hunting groups observed during the past century have sometimes practiced infanticide, it seems likely that this was also the case
in human prehistory.

It is also possible that sex-selective infanticide has been practiced
since early prehistoric times. Henry Vallois has studied the available
sexually identified human fossil remains from the Pithecanthropines
to the Mesolithic era, and found a sex ratio of 148.[13] Such a high ratio
cannot readily be explained by the early deaths of many women in
childbirth, since there is no apparent reason why the bones of such
women should have been less apt to have become fossilized than those
of men. There are other possible explanations, such as an undetected
bias in the sexual identification of human fossil remains. It may be,
however, that the apparently high sex ratios were due to female
infanticide. We can only speculate about the reasons why such a
practice might have existed in those presumably prepatriarchal
times. Perhaps the hunting skills of males were more highly valued
than the gathering skills of females, even though the latter may have
provided a higher proportion of the food consumed. Or perhaps
patriarchal institutions have a longer history than we have any way of
knowing.

We do know, however, that patriarchal social institutions make
sex-selective infanticide much more probable. In societies which are
matrilineal and/or matrilocal, and/or in which women enjoy a
relatively high degree of autonomy and social influence, daughters
are far less likely to be killed because of their sex.[14] There are very few
cultures in which male infants are more apt to be killed than females.

One such culture is that of the Mundugumor of Papua New
Guinea, as described by Margaret Mead.[15] The reason for daughter-
preference among the Mundugumor is not female dominance, for
they are not a female-dominated society, although the women are the

primary producers of food, and are very nearly as competitive and aggressive as the men. By a seeming paradox, it is Mundugumor fathers who value daughters more than sons, while mothers tend to value sons more highly. Mead explains that fathers value daughters more for three reasons: (1) according to the "rope" system of lineage, daughters inherit from their fathers and sons from their mothers; (2) women, as economic producers, are a source of wealth; and (3) fathers can exchange their daughters for additional wives, whereas sons will compete with fathers for the privilege of exchanging their sisters.

There have been many societies in which female infanticide has been so common that adult women have become scarce and wives have had to be captured from other groups. The Scottish anthropologist John McLennan hypothesized that female infanticide and "marriage by capture" were the key parts of the process by which patriliny and patrilocality replaced the matrilineal and matrilocal family structures which, he argued, must have originally existed in all human societies.[16]

Female infanticide and the capture of adult women were a common pattern in the Arab world from prehistoric times to the nineteenth century.[17] Fathers often ordered that female infants be buried alive immediately after birth. At times, the social pressure to kill female children was so great that parents did not dare to save daughters even when they were their only children.[18] The birth of a daughter was regarded by Arab men as a humiliating calamity—and often still is.[19] The pre-Islamic attitude towards female infanticide was expressed by a proverb to the effect that to bury a daughter is a generous act, because there will be one less mouth to feed.[20] Mohammed prohibited female infanticide, but it is only in this century that it has become relatively uncommon in Islamic societies. Poverty was a major reason for female infanticide among the Arabs; raising daughters was an unaffordable expense. Another reason frequently cited was the fear of dishonor should a daughter be captured in war.

Female infanticide was also common in traditional China. Male children were valued as support for the parents' old age and because they would inherit the family name and continue the worship of the family ancestors. Female children were far less welcome, and were often exposed or given to other families to be reared as wives for the latters' sons. These adopted daughters were often neglected or ill-treated, and not infrequently died as a result.[21] Jesuit missionaries reported in the seventeenth century that in Beijing, "several thousand babes (almost exclusively female) were thrown on the streets like refuse, to be collected each morning by carriers who dumped them into a large pit outside the city."[22]

This custom seems to have persisted at least until the 1830s,[23] and female infanticide was practiced openly and extensively for at least another half century after that. In the 1870's Adele Fielde, a missionary and naturalist working in the Swatow region of China, sought to determine the extent of female infanticide. She interviewed forty women, who had borne a total of 183 sons and 175 daughters. Of the sons, 126 had lived to be over ten years old, while only 53 of the daughters had survived to that age; 78 of the daughers had been deliberately destroyed.[24] Since the institution of the one-child policy in China, it has been reported that the sex ratio among young children is again extremely high. In one city there have been said to be five times as many male children under five years of age as females.[25]

In India, many tribes practiced female infanticide from prehistoric times until the latter part of the nineteenth century, when through the efforts of British colonial rulers and native reformers its frequency was finally reduced.[26] The Rajputs, Jharejas, and certain other tribes in northern India killed virtually all female infants at birth.[27] Female infanticide was "so much a *mechanical conduct* on the part of the Jharejas that ... mothers did not very often bother to appraise Jhareja fathers [of] any birth of [a] daughter."[28] Far from considering the elimination of female infants a crime, the Jharejas regarded it essential to their pride and social status, and vigorously resisted British efforts to suppress it.[29]

The expense of providing a daughter with a dowry is another motivation for female infanticide in India. The provision of cash dowries was outlawed in 1961, but the custom persists through the giving of material goods. A girl cannot hope to marry well without a sizable dowry, and the groom's family often makes exorbitant demands. India is currently experiencing an epidemic of "dowry murders." Thousands of young women have been burned to death by their husbands or in-laws, as revenge for a supposedly inadequate dowry.[30] The birth of a daughter is still mourned as a disaster, while the birth of a son is celebrated by the families of both parents.[31] Female children often receive less food, less affection and inferior medical care, with the result that infant mortality remains considerably higher for females.[32] The average life span of adult males is also longer than that of females.[33]

Infanticide has been no less common in western than in eastern civilizations. Sarah Pomeroy's examination of ancient burial records suggests an extremely high sex ratio in archaic, classical and Hellenistic Greece, and in Rome and most of the Roman Empire. Burials from archaic Greece show a ratio of two males to one female. Homer says that King Priam had 50 sons but only 12 daughters.[34] In Pomeroy's view, the burial evidence, like the literature and mythology, strongly suggests a high rate of female infanticide. Some critics have argued that the most probable explanation for the high sex ratios found in ancient burials is that fewer women were given burials of a sort which would enable their remains to be discovered by archeologists. But there is little evidence that this was the case. The scepticism about Pomeroy's findings may largely due to reluctance to believe that the approximately equal sex ratios to which we are accustomed may not be typical of all times and places.

The evidence for female infanticide in Greek societies during the third and second centuries B.C. is particularly clear. Sir William Tarn found that in 79 families of which details were recorded, there were 118 sons and only 28 daughters.[35] The Jews, who prohibited infanticide very early in their history, were unusual in this respect. Among other groups, infanticide was almost never made illegal until the Roman Empire adopted Christianity as its official creed in the fourth century A.D. It was not regarded as a moral issue, and physicians found it natural to discuss the criteria for determining which newborns were worth rearing.[36] It went without saying that boys were more often worth rearing than girls. A first century B.C. husband wrote to his wife, "If, as may well happen, you give birth to a child, if it is a boy let it live; if it is a girl, expose it."[37]

Scholars have also begun to study the prevalence of infanticide in Christian Europe. Because infanticide was prohibited by church and (sometimes) civil authorities, direct evidence is difficult to find. However, any marked imbalance in favor of young males must be taken as presumptive evidence of the selective killing or neglect of female children. Such imbalances have been found in many instances. Josiah Russell found artificially high sex ratios among the survivors listed in the *Inquisitions Post Mortem* in thirteenth through sixteenth century England.[38] Emily Coleman studied demographic data from

the Saint Germain-des-Pres region of France during the same period.
She found that the larger the family, the higher the sex ratio. Smaller
farms and poorer and more heavily populated regions also exhibited
very high sex ratios—as high as four male children to one female.[39]
Coleman concludes that female infanticide must have been a standard
means of eliminating economically unsupportable mouths.

The incidence of infanticide in Europe did not begin to abate until
the present century, with the greater availability of contraception and
medically safe abortion. The killing or abandonment of newborn
infants was so common in eighteenth- and nineteenth-century Europe
that it produced an extensive literature and numerous reform efforts.
The 1890 edition of the *Encyclopedia Britannica* stated that, "The
modern crime of infanticide shows no symptoms of diminution in the
leading nations of Europe."[40] Foundling hospitals were operated in
many European cities as early as the eighth century, but "The
chronicle of the hospital everywhere was one of devoted effort but
unrelieved tragedy."[41] The majority of infants left in the care of such
institutions died of malnutrition or disease. Elizabeth Badinter
reports that in France,

> In the last third of the eighteenth century, the percentage of children who
> died after being left at a hospital was more than 90 percent at Rouen, 84
> percent in Paris, and 50 percent in Marseilles.[42]

The survival rates of female infants in the foundling hospitals were
sometimes higher than those of males. Yet a larger number of girls
died there because more were abandoned in the first place and fewer
were later reclaimed by their parents. Richard Trexler, in a study of
sixteenth-century Florentine records, found that a significantly higher
proportion of infant girls died not only in the hospitals, but also in the
care of wet nurses, or *balie*. Female infants were also more apt to die of
suffocation or "overlaying," which was a suspiciously common cause
of infant death.[43] Trexler notes that: "In law, in the family, and in the
foundling home, European society preferred boys. This meant more
deaths for infant girls."[44]

In contrast to the open way in which infanticide was practiced in the
ancient world, in Christian Europe it was publicly condemned but
practiced covertly, in ways that made it appear accidental or
inadvertant. Not only were females the most frequent victims of
infanticide; adult women were also victimized by the penalties
imposed for infanticide. The Church condemned infanticide in the

strongest terms, and often imposed rather severe penances (e.g., a year on bread and water) upon married women whose infants died under suspicious circumstances. But there were very few civil prosecutions of married persons for the crime of infanticide: "How could one prove infanticide within the walls of the family home? Who would want to?"[45] In contrast, unmarried women who were suspected of having disposed of their infants were often executed in the cruelest manner which could be devised.

> Throughout the second millennium all over the European continent it was common practice to execute unwed mothers for killing their babies, regardless of the circumstances surrounding the child's conception or death. Such vengefulness was not questioned until the eighteenth century. The death penalty existed even in towns where there were no hospitals or other places for surplus children, in some where the mother literally had no place to go.[46]

The sixteenth century, which saw the torture and execution of hundreds of thousands of women (as well as many men) who were suspected of being witches, also saw unprecedented numbers of women executed for alleged infanticide. The Carolina, a criminal code promulgated throughout the Holy Roman Empire in 1532, decreed that all "infanticides" were to be buried alive, impaled, or drowned. At the same time, the concealment of pregnancy was made a crime, and unmarried women whose pregnancies had been concealed were presumed to be guilty of infanticide if the infant died. A common method of execution was by "sacking"; the woman was sewn into cloth sack together with a cat, a dog, and a viper or some other animal, and the sack was thrown into a river. One Saxon jurist named Benedict Carpzov (1596-1666) boasted of having executed 20,000 women for infanticide, witchcraft, or both.[47]

The extraordinarily severe punishment of unmarried women suspected of infanticide persisted into the nineteenth century. As Maria Piers points out, Christian Europe's ambiguous feelings about infanticide were resolved by a seemingly bizarre double standard. Murderous penalties were inflicted upon unmarried women whose infants died, while parents and wet nurses "were given the license to kill on a grand scale."[48] Unmarried mothers were made scapegoats for the society's bad conscience, while the deaths of "legitimate" infants were disguised as accidents, or blamed on witchcraft. Unmarried fathers, of course, were seldom held accountable in any way.

Wherever infanticide is a customary practice, it is the result of perceived social, economic, or other necessity. In most cultures not influenced by the Judeo-Christian religion, infanticide has not been regarded as a form of murder. People take it for granted that some newborns must die in order that older children and adults may survive. In severely patriarchal societies it has generally been considered a man's right to decide which of the children borne by his wife or female slaves should be reared. An infant was not regarded as having a right to life until certain ceremonies had been performed, usually within a few days of birth. In Babylonia, the father had to give the child a soul by blowing into its face and giving it the name of one of his ancestors.[49] Among the Arapesh of Papua New Guinea, the father decided whether or not the newborn infant should be washed. If washed, it became a member of the family; if not, it was exposed.[50] In Athens, a child was accepted into the family through the amphidroma ceremony, which

> was performed as a rule on the fifth day, when the new baby was carried by its nurse around the ancestral hearth to receive consecration and a name. If the child was not wanted, the father had to dispose of it before the amphidroma.[51]

Such ceremonies have enabled societies which practice infanticide to maintain a clear distinction between the killing of newborns and the murder of older persons. There is no reason to believe that the social acceptance of infanticide leads to a diminution of parental affection for surviving children or to a loss of respect for human life in general. The Arapesh, for example, in spite of the routine practice of infanticide, are among the most gentle and child-oriented people ever studied. Mead reports that both women and men regard the nurturing of children as their greatest pleasure and finest achievement. Infanticide is by no means the *only* way that people lacking effective contraceptives and safe means of abortion can maintain the necessary balance between the human population and available resources. It is, however, a means of population control which most cultures, prior to the present century, have found necessary. As such, it cannot be condemned unless it can be shown that a realistic and morally preferable option was available—which very often was not the case.

Nevertheless, systematic sex-selective infanticide must be regarded as morally objectionable, and thus a form of gendercide. Sex-neutral infanticide is often a measure made necessary by conditions beyond the control of either the individual parents or their society. But

sex-selective infanticide is a consequence not just of economic necessity, but also of sexist attitudes and institutions. It is morally
objectionable not only because it reflects the injustices of patriarchal
social systems, but because it contributes to the perpetuation of those
injustices. When infants are deliberately killed or allowed to die
simply because they are female (and when it is common knowledge
that this occurs), all females are dramatically reminded of the social
devaluation of their sex. As married women they will pray for sons,
knowing that should they bear daughters the latter's lives may be
forfeit, and their own value diminished. Whatever status they may
gain as the mothers of sons will be contingent upon the society's
underlying contempt for females.

To say that female infanticide is a form of gendercide is not to
condemn people who have practiced it as morally benighted. The
responsibility rests with the patriarchal system rather than the
individual parents. Female infanticide and selective neglect occur
because the objective circumstances created by patriarchy lead
parents to conclude that they simply cannot afford to raise daughters,
or as many daughters as sons.

The cruel punishments inflicted in Christian Europe upon unmarried mothers whose infants failed to survive were also a form of
gendercide. Like the witch persecutions with which they were associated, they were due in large part to an abhorrence of female sexuality
outside the context of patriarchal marriage. Had they been based
upon a genuine respect for the lives of infants, married women (and
all men) would not have been virtually exempt from civil punishment
for the crime of infanticide.

The European Witch Hunt

The number of persons executed during the long European witch
craze cannot be known.[52] Estimates range from the hundreds of
thousands[53] to over nine million.[54] The belief in some form of
witchcraft is a very ancient one, found throughout most of the world.
Occasional persecutions of suspected witches can be found in most
periods of history. But the most massive witch hunt of all began in the
fourteenth century. It eventually spread across most of Western
Europe, creating a reign of terror which lasted for roughly four
centuries. In France, Belgium, Germany, Poland, Switzerland,

northern Italy, England and Scotland, droves of unfortunate individuals suspected of witchcraft were tortured, often in sexually humiliating ways, before being burned, drowned, or hanged.[55] Not all of the supposed witches were women, but the great majority of them were. Widowed or unmarried women seem to have constituted the largest proportion of the victims.

Witches were believed to have placed themselves under the influence of the devil, in return for the power to use magic to do evil and sometimes good. They were thought to have the power of flight, which they used to attend Sabbats—diabolical congregations at which they copulated with devils, slaughtered babies, ate human flesh, and did other atrocious things. These beliefs were not merely the superstitions of uneducated peasants; they were fully accepted as scientific fact—at least so far as public utterances went—by virtually every educated man and woman in Europe.[56] Numerous learned treatises were written explaining the nature of witches and their doings, ways of recognizing witches, and procedures for wringing confessions from suspected witches through torture and intimidation.

There is no generally accepted explanation of why the large-scale persecutions of alleged witches began when they did, why they continued for so long, or why the slaughter was so great. Yet there is little mystery about why the great majority of the victims were women. The blame must attach to the misogynist ideas inherent in the Greek and Judeo-Christian traditions.

In Greek mythology, evil is said to have entered the world through the first woman. Pandora was sent by the gods to punish men for their pride. She was given a box which she was told never to open, but she opened it anyway and thereby released all the forms of human suffering. In Hebrew and Christian mythology, woman is held responsible for humanity's fall from divine grace. Death, suffering and sin were visited upon humanity because Eve, like Pandora, could not resist the lure of forbidden fruit. For her disobedience, Eve was cursed by God, who condemned all women to bear children in pain, to desire their husbands, and to be ruled by them.[57] Christianity intensified this tradition of misogyny through its condemnation of sexuality as inherently defiling, and of women as lustful temptresses. The Church Fathers literally regarded all women as agents of the devil. Tertullian told women that, "God's sentence hangs still over all your sex and his punishment weighs down upon you. You are the devil's gateway."[58]

Throughout European history, witches are generally depicted as women—usually old women, but sometimes young and seductive ones. For several centuries prior to the great witch hunts, the Church denied the reality of witches and declared the belief in witches a heresy. Thomas Aquinas helped to reinstate the official belief in demonology. Aquinas argued that women are more apt to become witches than men because of their moral and intellectual weakness. But the "science" of demonology did not reach its full flowering until the fifteenth century, with the writings of Torquenada and other agents of the Inquisition. For three centuries the Bible of the witch hunters was the *Malleus Maleficarum*, which was published in 1486 by two Dominican Inquisitors. Having recently condemned to death a large number of accused witches, most of whom were women, Kramer and Sprenger were at pains to explain why witches were usually women. Their explanation was that: "All witchcraft comes from carnal lust, which is in women insatiable ... Wherefore for the sake of fulfilling their lusts they consort with devils."[59]

Such attitudes towards women and female sexuality are largely a result of the sexual double standard associated with patriarchal marriage. Because it gives men the right to control women's sexual and reproductive lives, patriarchal marriage makes any exercise of sexual autonomy by women appear threatening and even diabolical. From there it is a short step to the conclusion that women themselves are sinful, deceitful, and potentially in league with supernatural forces of evil. Widows and unmarried women who are not under direct patriarchal authority are particularly apt to be seen as a threat to the social order.

It has been suggested that the European witch hunts may have been triggered in part by an increase in the number of such women living outside direct male control. Such an increase may have resulted from a higher death rate of men during epidemics of bubonic plague, and/or a pattern of later marriage, with more people remaining permanently unmarried.[60] In the words of Kramer and Sprenger, "When a woman thinks alone, she thinks evil."[61] The multitudes of women who were murdered because they were suspected of thinking alone were victims of gendercide.[62]

Suttee, or Widow-Sacrifice

It was once a common practice in many parts of the world to kill the
wives, concubines, and female (and sometimes male) servants of
deceased patriarchs, perhaps in the hope that they would continue to
serve their masters in the next world. Widow-sacrifice has existed not
only in India, but in Scandanavia, Slavia, Greece, Thrace, Scythia,
Ancient Egypt, among the Tongans, Balinese, Fijiians and Maoris, in
some African and Native American tribes, and in China.[63] India is
unusual only in that there the practice has continued into modern
times.

The historical origins of suttee are unknown. It was already practiced in Punjab in the fourth century B.C., when Alexander's soldiers
brought back accounts of it. The practice did not exist among the
ancient Indo-Aryans, and was not sanctioned in the original Vedas,
the Hindu sacred scriptures. However, these were apparently altered
in the first and second centuries A.D. to condone the ritual sacrifice of
widows.[64] The word "suttee" is an anglicized version of the name
"Sati." Sati was the wife of the god Shiva, and was said to have killed
herself because of an insult to her husband. Her name came to refer
first to the devoted wife who never remarried after her husband's
death, and then to the widow who burned herself to death on her
husband's funeral pyre. This was originally an upper-class practice,
but often tended to spread downwards.

Apologists for suttee have maintained that it is a noble act of
devotion on the woman's part, performed proudly and without
coercion. Such, indeed, were the images with which it was surroun-
ded. The sati was thought to gain the highest form of salvation, and to
have the power to relay prayers directly to the gods. Many of the
shrines erected in honor of satis still exist. Their deaths were often
voluntary in the minimal sense that no visible means of coercion were
employed. Some widows even insisted upon burning themselves, in
spite of their families' attempts to dissuade them. But often there was
not even the pretense of voluntariness. Accounts written by Moslem
and European observers tell of the wives and concubines of wealthy
chieftains "being dragged resisting to the great pyre and forcibly
consigned to the flames."[65] To prevent escape, the pyre was sometimes laid in a pit; or the woman was chained to the logs or pinned
beneath them. Sometimes, after the death of an important man,

hundreds or even thousands of women were burned. Rajput women were sometimes burned *before* they became widows. When the men felt that military defeat was inevitable, they sometimes rounded up all of the women and children and burned them alive, in order to prevent their falling into the hands of the enemy.

Suttee was banned in the British-ruled parts of India in 1829, but persisted for many decades in northern and western India, and in Nepal, Assam, and Bali. It has never entirely disappeared, and cases continue to be reported each year. In 1980, a sixteen-year-old widow burned herself on her husband's pyre in the Indian village of Jhadi. The site has since become a place of pilgrimage where the sick hope to obtain miraculous cures.[66]

Mary Daly has described suttee as "the ultimate consummation of [patriarchal] marriage."[67] The Hindu ideal of the virtuous woman is the slavishly devoted wife whose relationship to her husband is as that of a mortal human being to a god. Although girls are sometimes married while still children, sometimes to men old enough to be their grandfathers, it is traditionally regarded as shameful for a woman to remain alive after the death of her husband. Tradition holds widows responsible for their husbands' deaths; if they have done nothing wrong in their present life they are presumed to have misbehaved in some previous life. Widows who choose to survive may still be subjected to lifelong humiliation and abuse. The widow's family often insisted upon suttee for the sake of honor (i.e., to eliminate the risk of sexual misbehavior or remarriage by the widow), and to eliminate one of the husband's heirs. But the underlying cause is the misogyny and contempt for women which are inevitable in severely patriarchal societies.

Genital Mutilation

The mutilation of women's genitals is also a form of gendercide, even though the resulting deaths are peripheral to the avowed purposes of the practice. In the past, various forms of clitoridectomy and infibulation have been practiced in many parts of the world. The Romans infibulated slave women by fastening their labia together with iron rings, and husbands in Christian Europe sometimes used the same method to safeguard their wives' virtue. Iron chastity belts served the same purpose, and caused at least as much suffering. Such devices were openly sold in Europe up to the end of the nineteenth century.[68] Women are said to have been infibulated in some tribal groups in Central and South America,[69] and in precolonial Australia.[70] In the United States, clitoridectomies were performed for the treatment of "hysteria" and many other supposed mental or physical ailments. This practice persisted into the 1930s, and possibly even later.

The form of mutilation which is known as female circumcision existed in Africa long before the arrival of the Islamic religion. However, its survival in much of Africa today is largely due to Islamic influence. Although the Koran does not require that women be circumcised, it is a common belief that the Islamic faith does require it, and few Islamic leaders have spoken out against it.[71] It is widely practiced in Egypt, Somalia, Nigeria, the Sudan, Kenya, Ethiopia, Mali, Sierra Leone, the Ivory Coast, and some other parts of Africa.

The term "female circumcision" gives the impression that the procedure is roughly equivalent to the circumcision operation performed on men. But whereas male circumcision usually involves only the removal of part or all of the foreskin of the penis, leaving the organ largely intact and functional, female circumcision often involves the complete excision of the clitoris, labia major and labia minora—i.e., all of the external genitalia. This is known as Pharaonic circumcision, because it is thought to have been performed on women in ancient Egypt. Less severe forms are also practiced, but the extreme Pharaonic form is still the most common.[72]

These mutilations are performed by force on girls usually aged 8 to 12, often by a midwife or by the girl's female relatives. Anaesthetics are usually not used, and the conditions are often unsanitary in the extreme. The edges of the wound are stitched together, and the girl is forced to lie with her legs tied together until it has healed, leaving only a tiny opening for passing urine and menstrual blood. (Some hospitals now perform the mutilation using modern medical procedures, but this can hardly be viewed as a reform.) The immediate complications may include shock, fever, infection, retention of urine and menstrual blood, difficult or painful urination, damage to the bladder or rectum, and sometimes death.[73] The long-term complications include pain during intercourse, inability to experience sexual pleasure, chronic or recurrent infection, inability to control urination or to urinate without pain, and a variety of difficulties in pregnancy and childbirth.[74] The infibulated woman must be cut open for each birth, and sewn up again afterwards. Often her husband can "de-flower" her only with a knife, after which intercourse takes place without giving the wound time to heal.

There are no statistics available on the number of girls and women who have died as a result of genital mutilation. Because it is illegal in many places, it is generally done in secret, and if death results the cause is likely to be disguised. The long-term complications may be a contributing cause of death at any later point in the woman's life, but she may be unaware that her problems are a result of her previous mutilation.

Like suttee and witch hunting, genital mutilation serves not only to eliminate some women, but to control the behavior of all women. It is usually women who perform the operation, but it is ultimately men who demand it. It is said that no man will marry an uncircumcised woman;[75] and marriage remains women's primary means of survival in these severely patriarchal societies.

Why should men demand genitally mutilated wives? Many can give no reason other than that it is customary. Some African leaders have defended the practice as an important part of the people's culture and religion, and the United Nations has been reluctant to investigate or oppose it for this reason. Beauty, cleanliness, and increased sexual pleasure for the husband are sometimes mentioned as reasons, as is the belief that it is a requirement of the Islamic religion.[76] But the primary function can only be to eliminate women's capacity for sexual pleasure and thus ensure their sexual purity. "Son of an uncircumcised women" is regarded as one of the vilest possible insults; for an unmutilated woman is presumed to be promiscuous or in danger of becoming so.

The Denial of Reproductive Freedom

Not all patriarchal societies have practiced such dramatic forms of gendercide as witch burning, suttee, and genital mutilation. But in every patriarchal society countless women have died as a result of the denial of sexual and reproductive freedom. In most patriarchal systems, women are not only denied the right to voluntary extramarital sexual activities; they are also denied the right to *refuse* sexual intercourse with their lawful husbands. Furthermore, they are often denied effective means of resisting the unwanted sexual attentions of men not their husbands. This sexual enslavement, which is often combined with the deliberate suppression of existing means of contraception and medically safe abortion, has probably caused the deaths of more adult women than any other form of gendercide. The World Health Organization estimates that 200,000 women around the world die annually from illegal abortions.[77]

Women from prehistoric times to the present century have suffered a high death rate from childbirth and its complications. The evolution of the large human brain has made giving birth far more difficult and dangerous for women than it is for most female mammals. Women who are healthy, active, and free to space their pregnancies several years apart may give birth more easily than those who are poorly nourished, confined to a sedentary life, or subjected to constant pregnancy. But prior to the development of modern obstetrics, childbirth has always involved a significant risk of death. A study of Neanderthal remains found that men at age 20 had a life expectancy 40 percent longer than that of women.[78] Whereas males often have a higher natural death rate in infancy, women have usually had a far higher mortality rate during their fertile years. This remains true in many of the impoverished parts of the world.

In Europe the excess mortality of women during their childbearing years has disappeared only in the present century, thanks not only to safer obstetric methods but also to improved means of contraception and abortion. According to Edward Shorter,

> Before 1800 perhaps 1 percent to 1.5 percent of all births ended in the mother's death. The exact average of the studies ... consulted is 1.3 percent ... If we assume that the typical woman who lived to the end of her fertile years gave birth to an average of six children, her lifetime chances of dying in childbirth would be 6 times 1.3, or 8 percent.[79]

It would be a mistake to suppose that the high maternal death rates prior to the present century were simply the result of primitive obstetric methods and the lack of reliable contraceptives. Patriarchy has increased the maternal death rate by forcing girls to begin childbearing before their bodies are mature, by depriving them of adequate nutrition, by suppressing knowledge and access to contraception, abortion and midwifery, and above all by requiring women to be sexually available to men. Margaret Sanger, who led the fight for legal contraception in America, described the plight of women who had no means of preventing pregnancy, even though they knew that they could not survive another birth. Financial dependence kept them tied to men who literally cared more for their own pleasure than for their partners' lives.[80]

The "right to rape" which patriarchy accords to husbands is by no means a thing of the past. In most American and other states, rape is legally defined as forcing sexual intercourse upon a woman *not one's wife*. Some states have reformulated the definition to include "spousal rape." But prosecutions of rapes by legal or de facto husbands are still rare enough to attract headlines—not because such rapes are uncommon, but because convictions are exceedingly difficult to obtain.

The persecution of female midwives and the virtual elimination of their trade in the industrialized nations is another gendercidal aspect of the denial of reproductive freedom. When the male medical profession first took over the practice of midwifery, gradually driving out female practitioners, the most noticeable result was an enormous increase in maternal deaths.[81] (See Chapter 6, fourth section.) During the nineteenth and early twentieth century, midwifery was outlawed in most of the American states, leaving many women who could not afford to pay the fees of male physicians without medical assistance in childbirth. Some midwives were guilty of dangerous and unsanitary practices; but many had skills and knowledge which could have saved a great many lives had the male physicians not assumed that they could not possibly learn anything from female practitioners.

Female midwives were persecuted long before the male medical profession learned that there was profit to be made from obstetrics. During the era of witch hunts, midwives and women who provided advice on contraception and abortion were often executed as witches. The Church frowned upon efforts to make childbirth safer or less painful, because it held that God had condemned women to bear children in pain.

Thus, over the millennia, millions of women have died as a result of pregnancies which patriarchal institutions prevented them from either preventing or safely terminating. The high incidence of infanticide in almost all of the patriarchal civilizations of the past must also be blamed in large part on the denial of women's reproductive freedom. Women who are free to regulate their own reproductive lives are far less apt to bear unwanted children. Patriarchy cannot be excused of responsibility for these deaths on the grounds that women know what is required of them in marriage and should refuse if they do not like the terms. For it is a characteristic feature of patriarchy that male domination of the means of subsistence ensures that few women can survive—let alone raise children—without in one way or another selling themselves to men.

Misogynist Ideologies

Gendercide and female-denigrating ideologies are the twin offspring of patriarchy. Neither can be understood in isolation from the other. Misogyny is such a pervasive element in our own culture that it often goes unnoticed.

> The hate, fear, loathing, contempt, and greed that men express toward women so pervade the human atmosphere that we breathe them as casually as the city child breathes smog. ... Much of the work that feminists do is an effort to resensitize us to it: smog has to be identified as a problem before citizens can decide that they would prefer cleaner air.[82]

Misogyny lies at the heart of the Judeo-Christian and many other patriarchal religions. The biblical God is unquestionably male, although he is sometimes described as incorporating feminine elements. In the more familiar of the two creation stories related in *Genesis*, man is created first, and is said to be in the divine image. Eve is created for his sake, but proves to be a source of misery for all "mankind." Saint Paul reminds his Corinthian congregation that man

> is the image of God and reflects his glory; while woman is the reflection of man's glory. For man was not made from woman, but woman from man, and man was not created for woman, but woman was for man. That is why she ought to wear upon her head something to symbolize her subjection.[83]

The theme of female defectiveness also runs throughout western philosophical and scientific literature. Plato, although he advocated the admission of women to all responsible public roles, nevertheless assumed without argument that "females are inferior in goodness to males;"[84] and consistently used the adjective "feminine" as an expression of contempt. Aristotle defined the female as "a deformed male,"[85] deficient in the essentially human element of soul, intellect, or rationality, and therefore naturally subject to male rule. In his view, women contribute nothing to the process of reproduction except matter; the rational soul of the child comes from the father alone. Following Aristotle, Aquinas held that, woman is defective and misbegotten; for the active force in the male seed tends to the production of a perfect likeness of the masculine sex; while the production of a woman comes from defect in the active force or some material disposition.[86]

There is no need to multiply examples of this sort. It is sufficient to note that Kant, Locke, Rousseau, Hegel, Darwin, Freud, Jung, and thousands of lesser-known male thinkers have presented versions of the view that women are morally and intellectually defective and must therefore be dominated by men.[87] There has not, until quite recently, been any parallel body of literature produced by women and designed to prove male inferiority. Thousands of years of enforced silence lie behind the furious response of some contemporary feminists, such as Valerie Solanas, who has described the male as "an incomplete female, a walking abortion,"[88] and a "subhuman animal."[89]

Not all men consciously regard women as their inferiors. But all of us are participants in ancient cultural traditions which make sacred or scientific dogmas out of uncritically androcentric sentiments. One way to glimpse the misogyny which still pervades our own culture is to look at the degraded images of women presented in the more sadistic forms of contemporary pornography, which depict the rape, torture, mutilation, or murder of women for the gratification of male viewers. Such material glorifies anti-female gendercide.

Another way to glimpse this pervasive misogyny is to consider the "four-letter words" which are still sometimes used to refer to women—particularly by men and in the company of other men. It is significant that some men still refer to women by derogatory terms relating to female sexual organs. As Robert Baker points out, such "metaphorical identifications" reveal underlying attitudes towards the person or object referred to.[90] These terms reflect not just a sexual interest in females (which, of course, is not necessarily a sign of misogyny), but the perception of women as simultaneously defined and degraded by their sexuality. They do not connote a fellow human being who happens to be sexually attractive, but a vaguely disgusting thing, to be used carelessly and with contempt.

The history of gendercide and misogyny which which we share with most other human societies is a highly relevant part of the social context into which the new technologies of sex selection are being introduced. The systematic devaluation of women gives credence to the fear that the new methods of selection may function as yet another form of gendercide.

Notes

1. Andrea Dworkin, *Our Blood: Prophecies and Discourse on Sexual Politics* (New York: Harper and Row, 1970); and *Woman Hating* (New York: E. P. Dutton and Company, 1974).

2. Mary Daly, *Gyn/Ecology: The Metaethics of Radical Feminism* (Boston: Beacon Press, 1978).

3. For instance, infants who were born feet-first might be considered unlucky and therefore killed, as was reported to be the practice among the Kgatla of South Africa; see I. Schapera, *Married Life in an African Tribe* (London: Faber and Faber, 1911), 225.

4. This was the case among many of the Australian aborigines (see W. G. Sumner, *Folkways* [Boston: Ginn and Company, 1906], 315), and various African tribes, such as the !Kung (Richard Borshay Lee, *The !Kung San: Men, Women and Work in a Foraging Society* (Cambridge, England: Cambridge University Press, 317–320).

5. David Bakan, *Slaughter of the Innocents* (San Francisco: Jossey-Bass, 1971), 30.

6. J.M. Orsmund, *Ancient Tahiti*; cited by C.F. Potter, "Infanticides," in M. Leach, editor, *Dictionary of Folklore, Mythology and Legend* Vol. I (New York: Funk and Wagnalls, 1949), 522–23.

7. Upendra Thakur, *An Introduction to Homicide in India, Ancient and Early Modern Period* (New Delhi: Abhinav Publications, 1977), 55.

8. Bakan, 29.

9. Julia O'Faolain and Laura Martenes, editors, *Not in God's Image* (New York: Harper & Row, 1973), 19.

10. Lee, loc. cit.

11. Charles Darwin, *The Origin of Species and the Descent of Man* (New York: Modern Library, 1936), 430.

12. Among primates, males often kill infants fathered by other males, apparently because this increases their chances of siring young themselves; see Sarah Blaffer Hrdy, *The Woman That Never Evolved* (Cambridge, Massachusetts: Harvard University Press, 1981), 36, 70, and 76–95.

13. Henry V. Vallois, "The Social Life of Early Man: The Evidence of Skeletons," in *Social Life of Early Man*, Sherwood L. Washburn, editor (London: Methuen and Company, 1962), 225; cited by Lloyd De Mause, *The History of Childhood* (New York: Harper and Row, 1974), 27.

14. Nancy E. Williamson, *Sons or Daughters: A Cross-Cultural Survey of Parental Preference* (Beverly Hills: Sage Publications, 1976), 106–11.

15. Margaret Mead, *Sex and Temperament in Three Primitive Societies* (New York: Morrow, 1963), 191–92.

16. John F. McLennan, *Primitive Marriage* (Chicago: University of Chicago Press, 1979).

17. Raphael Patoi, *Sex and the Family in the Bible and the Middle East* (New York: Doubleday, 1959), 135; cited by Adrienne Rich, *Of Woman Born: Motherhood as Experience and Institution* (New York: W. W. Norton and Company, 1976), 120.

18. W. Robertson Smith, *Kinship and Marriage in Early Arabia* (the Netherlands: Anthropological Publications, Oosterhout N.B., 1966), 193.

19. V. R. and L. Bevan Jones, *Woman in Islam* (Lucknow: Lucknow Publishing House, 1941), 4.

20. Jones, 14.

21. Arthur P. Wolf and Chien-Shen Huang, *Marriage and Adoption in China: 1845-1945* (California: Stanford University Press, 1980), 236.

22. William L. Langer, "Infanticide: A Historical Survey," *History of Childhood Quarterly* 1 (1973-1974), 354.

23. John B. Beck, "On Infanticide in Relation to Medical Jurisprudence,"

Researches in Medicine and Medical Jurisprudence (New York, 1835); cited by Langer, 326.

24. Adele M. Fielde, *Pagoda Shadows: Studies from Life in China* (Boston: Corthell, 1884), 23–25; cited by Wolf and Huang, 230.

25. Jeremy Cherfas and John Gribbin, *The Redundant Male: Is Sex Irrelevant in the Modern World?* (London: The Bodley Head, 1984), 110.

26. Kanti B. Pakrasai, *Female Infanticide in India* (Calcutta: Editions India, 1970), 14.

27. Pakrasai, 12.

28. Pakrasai, 44.

29. Pakrasai, 109.

30. *Time*, July 4, 1983, 27.

31. Susila Mehta, *Revolution and the Status of Women in India* (New Delhi: Metropolitan, 1982), 207.

32. In 1969, female infant mortality in India was reported at 148.1 per 1000 live births, as against 132.3 for boys. (Mehta, 209.)

33. Mehta, 239.

34. Sarah B. Pomeroy, *Goddesses, Whores, Wives and Slaves* (New York: Schocken Books, 1975), 46.

35. Sir William Tarn, *Hellenistic Civilization* (London: Edward Arnold Ltd., 1959), 101.

36. Robert Etienne, "Ancient Medical Conscience and the Life of Children," *Journal of Psychohistory* 4 (1976–1977), 131.

37. DeMause, 25–26.

38. Josiah C. Russell, *British Medieval Population* (Albuquerque: University of New Mexico Press, 1948) 147–149.

39. Emily Coleman, "Infanticide in the Early Middle Ages," in S.M. Stuart, editor, *Women in Medieval Society* (Philadelphia: University of Pennsylvania Press, 1976).

40. *Encyclopedia Britannica*, Ninth Edition, 1890, 42.

41. Langer, 358.

42. Elizabeth Badinter, *Mother Love: Myth and Reality* (New York: Macmillan, 1980), 112.

43. Richard Trexler, "Infanticide in Florence: New Sources and First Results," *History of Childhood Quarterly*, I (1973-1974), 102.

44. Trexler, 110.

45. Trexler, 105.

46. Maria W. Piers, *Infanticide* (New York: W.W. Norton and Company, 1978), 45–46.

47. Bakan, pp. 37–38; Piers, 45–46.

48. Piers, 51.

49. Bakan, 32.

50. Mead, 33.

51. Bakan, 32.

52. Witch hunts also occurred in the American colonies, those in Salem, Massachusetts being the most infamous; however, the number of victims was very small in comparison with the European witch hunts.

53. See, for instance, Christina Larner, *Enemies of God: The Witch Hunt in Scotland* (London: Chatto and Windus, 1981), 15; Jeffrey Burton Russell, *Witchcraft in the Middle Ages* (Ithaca: Cornell University Press, 1972), 39; and Nancy van Vuuren, *The Subversion of Women, as Practiced by Churches, Witch-Hunters, and Other Sexists* (Philadelphia: Westminster Press, 1973), 71.

54. Matilda Joslyn Gage, *Women, Church and State* (New York: The Truth Seeker Company, 1893), 247.

55. Hanging was the method of execution in England and Scotland, where the

witch craze began later and never reached the proportions that it took on much of the Continent.

56. H.R. Trevor-Roper, *The European Witch-Craze of the 16th and 17th Centuries* (Middlesex, England: Penguin Books, 1969).

57. *Genesis* 3:16.

58. Quoted by Brian Easlea, *Witch Hunting, Magic and the New Philosophy* (New Jersey: Humanities Press, 1980), 34.

59. Heinrich Kramer and Jacob Sprenger, *Malleus Maleficarum*, translated by M. Summers (London: Puskin Press, 1949), 16–17.

60. H.C. Eric Midelfort, *Witch-Hunting in Southwestern Germany, 1562-1684* (Stanford, Calif.: Stanford University Press, 1972), 183–85.

61. Kramer and Sprenger, 43.

62. If, as I have claimed, the witch hunts were a form of anti-female gendercide, it may seem odd that many men were also executed as witches. Some of these were relatives of accused women, some were victims of vengeful accusations, and some were probably common criminals. Many others were suspected heretics. The witch hunts grew in part from the Church's war against unorthodox ideas—although they were as vigorously prosecuted by Protestant as by Catholic authorities. E. William Monter points out that the highest ratios of males among persons accused of witchcraft occurred in those regions where heresy and witchcraft were most closely linked. *Witchcraft in France and Switzerland: The Borderlands During the Reformation*, (Ithaca: Cornell University Press, 1976), 23.

63. Edward Thompson, *Suttee: A Historical and Philosophical Enquiry into the Hindu Rite of Widow Burning* (London: George Allen and Unwin, 1928), 24–25.

64. Benjamin Walker, *Hindu World: An Encyclopedic Survey of Hinduism*, Vol. II (London: George Allen and Unwin, 1969), 461.

65. Walker, 464.

66. Mehta, 207.

67. Daly, 113.

68. Elizabeth Gould Davis, *The First Sex* (Baltimore, Maryland: Penguin Books, 1972), 167.

69. Ashley Montagu, "Infibulation and Defibulation in the Old and New Worlds," *American Anthropology* 47 (1945), 464.

70. Ashley Montagu, "Ritual Mutilation Among Primitive Peoples," *Ciba Symposia* 8:7 (1946), 421–36.

71. Fran P. Hosken, *The Hosken Report: Genital and Sexual Mutilation of Females* (Lexington, Massachusetts: Women's International Network News, 1979), "History of Genital Sexual Mutilation," 8.

72. Asma El Dareer, *Woman, Why Do You Weep? Circumcision and Its Consequences* (London: Zed Press, 1982), 1.

73. El Dareer, 30–35.

74. El Dareer, 35–39.

75. El Dareer, 73–74.

76. Hoskin, op. cit., "The Reasons Given," 5.

77. Buffalo News, August 12, 1984; cited in *Free Inquiry* 4:1(Fall, 1984), 56.

78. C. Alsadi and J. Nemeskeri, *History of Human Life Span and Mortality* (New York: Macmillan, 1953), 184; quoted by Edward Shorter, *A History of Women's Bodies* (New York: Basic Books, 1982), 229.

79. Shorter, 230.

80. Margaret Sanger, *My Fight for Birth Control* (New York: Farrar and Rinehart, 1931).

81. See Barbara Ehrenreich and Deirdre English, *Witches, Midwives and Nurses: A*

History of Women Healers (Old Westbury, New York: The Feminist Press, 1973).

82.   Dorothy Dinnerstein, *The Mermaid and the Minotaur* (New York: Harper and Row, 1976), 88–89.

83.   Saint Paul, *Corinthians I*, 11: 7–10.

84.   Plato, *Laws*, translated by R.G. Bury (Cambridge: Harvard University Press, 1970), Book VI, 78 1B, 1–2.

85.   Aristotle, *Generation of Animals*, translated by A.L. Peck (Cambridge: Harvard University Press, 1943), 175.

86.   Thomas Aquinas, *Summa Theologica* (New York: Benziger Brothers, 1947); Vol I, 466 (Part I, Question 92, Article I).

87.   See Mary Anne Warren, *The Nature of Woman: An Encyclopedia and Guide to the Literature*, for expositions of these authors' views about female nature. (Inverness, California: Edgepress, 1980).

88.   Valerie Solanas, *SCUM Manifesto* (New York: Olympia Press, 1968), 3.

89.   Solanas, 32.

90.   Robert Baker, "Pricks and Chicks: A Plea for Persons," in *Philosophy and Women*, Sharon Bishop and Marjorie Weinzweig, editors (Belmont, California: Wadsworth, 1979), 23. The article is useful in spite of its unjustified tone of condescension towards women like Kate Millett who have explored other cultural manifestations of misogyny.

3

The Extreme Options

The new reproductive technologies invite speculation about extreme solutions to the problem of sexual inequality. Until now the absence of alternatives to the natural mode of reproduction has meant that the elimination of the division of the human race into two sexes could be contemplated only by writers of fantasy fiction. But now that such alternatives are on the verge of coming into existence, the maintenance of the sexual dichotomy may soon cease to be a precondition for the survival of our species. If both women and men continue to exist, and if we continue to reproduce via the union of ova and spermatazoa, it will no longer be because the only possible alternative is extinction.

The elimination of one sex or the other need not occur through any massive process of gendercide. It could conceivably come about through the voluntary choice of individual parents to have children of one sex only. At one extreme, the development of ectogenesis might make possible the complete elimination of women. Men might reproduce by cloning, using ova produced by isolated ovaries cultured in laboratories; or methods of reproduction might be developed which do not require the use of ova at all. At the opposite extreme, cloning, parthenogenesis or ovular merging might make possible the emergence of an all-female world. Artificial means of reproduction might even make possible the elimination of *both* sexes in favor of some androgynous or gynandrous form. Thus, sexual inequality might eventually be eliminated through the elimination of sexual differentiation itself. A more probable alternative is the formation of self-reproducing unisex communities.

Will Women Become Obsolete?

The phenomena of son-preference and anti-female gendercide might suggest that the total elimination of women is more probable than the total elimination of men. Patriarchal religions blame women for all of humanity's ills, implying that if only women could be eliminated and the species reproduced in some other way, the result would be an earthly paradise. All-male groups—priesthoods, armies, clubs, etc.—often resist the entry of women as desperately as though females were the carriers of a lethal plague. Every patriarchally socialized male learns to despise the "feminine" aspects of his own nature. Thomas Aquinas asked why God in his wisdom saw fit to create women.[1] Aquinas's answer was that women were created for one purpose only—to assist men in the work of generation. For in every other type of work, Aquinas argued, men can be more effectively assisted by other men.

Given the depth of male misogyny, what would stop men from outlawing the production of female children, should reproductive technologies advance to the point where women are no longer needed for reproduction? Such a move might be defended on grounds which are not overtly misogynist. It might be argued that either an all-female or an all-male society would be preferable to a society based on sexual inequality. Patriarchy has shown itself to have enormous staying power. The reforms of the past century have barely begun to undermine male domination of the human social world. Sexual inequality may be the oldest and most universal form of injustice. Radical feminists argue that it has been the model for all other forms of oppression and exploitation. Whether or not that is true, it is certainly true that patriarchy and male domination are destructive to human well-being. While there would undoubtedly be some injustices in a unisex society, it would no longer be possible to deny the fully human status of one half of our species simply because of biological gender. And perhaps if sexual inequality were eliminated it would become easier to overcome other forms of oppression.

But how could anyone who is not committed to some misogynist ideology possibly think that if one sex had to go it would be better to eliminate women? It is men who are threatening to destroy the world, with their wars, their nuclear weapons, and their irrational destruction of nature. The stereotypically "masculine" man is a callous, egotistical individual who does not hesitate to exploit others for his own benefit. The stereotypically "feminine" individual may be lacking in ambition, but she is also nurturant, compassionate, and gentle. Surely, if the world is to survive, it needs more individuals of the latter sort, not more of the former. But such qualities may have no direct relation to biological gender. If these stereotypical differences between women and men are merely the result of conditioning, social learning, the rearing of children by women, [2] or the sexual power imbalance within the patriarchal family, then perhaps a case might be made for the elimination of women as a form of eugenics. If female anatomy were to lose its reproductive utility, it might come to be seen as a disadvantage which ought not to be unnecessarily inflicted upon anyone. The male sex is in many ways biologically more privileged. The greater upper-body strength of the average male is valuable not just for combat or self-defense against other persons, but also for the imposition of his will upon the material world. In the words of Simone de Beauvoir, the body is

> the instrument of our grasp on the world, a limiting factor for our projects. Woman is weaker than man; she has less muscular strength, fewer red blood corpuscles, less lung capacity; she runs more slowly, can lift less heavy weights, can compete with men in hardly any sport; she cannot stand up to him in a fight... Her grasp in the world is thus more restricted.[3]

A further advantage of male physiology is freedom from the menstrual cycle, with its associated mood shifts and other physical and mental complications. As pregnancy becomes a less frequent occurrence in women's lives, menstruation becomes more so. Women who spend most of their reproductive years either pregnant or nursing menstruate far less often (perhaps only 10 percent as many times in a lifetime) than do women who bear no children or only one or two, and who nurse each child for at most a few months. Should pregnancy be superceded by artificial gestation, menstruation might become an even more otiose burden. The constant cycling of the permanently nulliparous women is, in biological terms, a useless drain on her life energy. If women of the future no longer need their wombs for gestation they might choose to have them removed, or to suppress the menstrual cycle through hormone treatments. But such medical interventions—which are already quite feasible—carry significant risks. Why, it might be asked, should women continue to be produced at all if they are no longer needed for reproduction? Why should we not give all of our descendants the advantages of male physiology?

Such a eugenic argument for the elimination of women would be invalid. There are compensatory biological advantages to being female. Females are less vulnerable to an enormous range of genetic diseases and defects than are males, apparently because of the extra genetic information carried on the second X-chromosome. Freed of the rigors of childbearing under primitive conditions, women live nearly a decade longer than men and are much less subject to heart disease and many other ailments. It seems that testosterone has the effect of shortening the male life span. Only males who are castrated early in life live as long, on the average, as females.[4] The relative value of greater muscular strength versus greater longevity and better health is partly a subjective matter, and partly a function of social conditions. In a violent society or one in which most work must be done by human muscle power, physical strength is enormously important. In a relatively nonviolent society in which machines are used to lighten the heavier forms of labor, it is less so. Even in a violent society, the disadvantages of lesser size and strength can be somewhat compensated by self-defense training. And perhaps even the menstrual cycle has some intrinsic value, e.g., as a pertinent reminder that we are a part of the natural world and subject to its laws. Besides, if medical science can find alternatives to the natural mode of gestation, it can surely find safer ways of eliminating menstruation than hysterectomy or hormone manipulation. That it has not already done so is probably due largely to the fact that so few biomedical researchers are women.

But even if it could be shown that male physiology is more desirable from an individual or social perspective, and even if women were to consent to their own extinction, there is little chance that the men of the future will choose to eliminate the female sex. In spite of the misogyny which pervades most patriarchal cultures, few men dream of creating an all-male world. All-male clubs are one thing, but a world without any women at all is quite another. There are very few idealized depictions of all-male societies in science fiction and utopian literature. Science fiction and fantasy shelves are filled with other variations on the theme of gender, from the wish-fulfilling hyperpatriarchal societies and matriarchal nightmares of anti-feminist authors to feminist visions of sexually egalitarian societies, benevolent matriarchies, and all-female utopias. But all-male utopias are conspicuous by their rarity. Those fictional works which depict a world in which women are extinct, or threatened with extinction, are anything but utopias. In novels such as Philip Wylie's *The Disappearance*[5] and Frank Herbert's *The White Plague*,[6] the demise of the female sex is portrayed not as ushering in a paradise on Earth, but as a catastrophe without precedent. If misogyny is such a powerful force in masculine culture and psychology, why are all-male utopian visions so rare?

The answer, of course, is that however much (some) men may hate and fear women, most men still value women—if not as friends, companions or lovers, then at least as servants, sex-objects or scapegoats. There is no automatic social mechanism which ensures that sex ratios will remain roughly even. Very high sex ratios have existed in the past and may exist again in the future. Yet, as de Beauvoir remarks, "No man would consent to be a woman, but every man wants women to exist."[7] In a world without women, men would have to assume responsibility for child rearing, and for all the mundane domestic tasks which make civilized life possible. The resistance which women everywhere encounter when they seek to persuade men to do their fair share of domestic labor demonstrates how little this prospect appeals to most men. In a world without women, the changeover to a homosexual erotic orientation might turn out to be surprisingly easy. But the loss of male privilege is something which few men would care to face. Masculinity would lose its symbolic value if there were no representatives of a contrasting and allegedly inferior state of being. Furthermore, there are many men who genuinely like women, and would regard the world as a poorer place without them, even if masculine privilege were to become a thing of the past.

Men who are not irrationally gynophobic can well afford to like women. Women have not stood between men and full membership in human society. They have not passed laws which denied men the right to vote, hold public office or enter the more lucrative professions. They have not made it impossible for men to walk the streets without fear of rape, or promulgated a sexual double standard which says that female nonmonogamy is good clean fun, while male nonmonogamy is to be punished with anything from contempt to death by stoning. They have not established dominant religions which say that men are defective beings who are responsible for all human misery. Instead, women have given men a degree of freedom which, while it has usually varied in accordance with economic class, has always been greater than that which women of the same class have enjoyed. It is because women have undertaken the work of child care and homemaking that men have been relatively free to take part in public life and nondomestic creative endeavors. Whether women have undertaken this work because of the kindness of their hearts or because they have had little choice is irrelevant. The point is that men would be fools to imagine that their lives would be better without women—even if asexual reproduction were cheap and easy. Male misogyny is probably an expression less of a genuine desire to be free of women than of a bad conscience stemming from unjust exploitation of women.

An equally important reason why women will probably survive the introduction of artificial means of gestation is that it is unlikely that these will prove economically feasible as a universal substitute for pregnancy. Ectogenesis would require the nine-month-long use of what are likely to be highly costly devices, plus (in all probability) the continual or frequent monitoring of fetal development by highly trained professionals. It could hardly be much less expensive than the intensive care of prematurely born infants, and might well be more so. It is possible that ways will be found to gestate human fetuses in the wombs of nonhuman animals, and that this will prove less expensive. But the moral and ideological objections to such a procedure are likely to prevent its general adoption.

Thus, there is little reason to fear that sex selection, together with ectogenesis, will lead to the total extinction of the female sex. It is not in men's interest to eliminate women, and even if it were, the economic resources necessary to replace women as childbearers with artificial gestation machines are unlikely to be available for a very long time.

An All-Female World?

The case for an all-female world may be somewhat more enticing. Expensive methods of ectogenesis would not be needed. Embryos could be conceived in vitro through cloning, ovular merging, or parthenogenesis, and then implanted in the womb to gestate naturally. Except for the single function of spermatogenesis, women do not need men either for individual survival or for the reproduction of the species. Mia Albright describes the asymmetrical dependence of men upon women as follows:

> The female sex uses, and produces what she uses, a fraction of the male body ..., and a fraction of the male sex. The male sex, on the other hand, requires the entire female body for nine months of gestation, plus the labor of childbirth, nursing, and rearing the male sex ... and the male sex uses massive numbers of womyn [sic] to reproduce the male sex which he cannot do for himself... We give birth to ourselves and to the men; men don't give birth to either themselves or us. We don't need man; he needs us.[8]

The elimination of men may soon be technologically feasible. In *The Redundant Male*, Jeremy Cherfas and John Gribbon point out that,

> The technology of asexual female reproduction in the human species really isn't that far off. If suitably dedicated women overcame any ethical objections and applied themselves to the task they could be cloned within the decade.[9]

Indeed, it is something of a puzzle why males should even have come into existence. Most one-celled organisms reproduce asexually, and this has not limited their survival capacity. Sexual reproduction greatly reduces a species' reproductive potential, since males contribute far less to the reproductive process. In most mammalian species, they contribute only their sperm, leaving the care of the young entirely to females. The usual explanation of the appearance and survival of bisexual species is that the mixing of the genes of different individuals in each generation increases the amount of genetic variation in the offspring, and thus speeds the evolutionary process. One-celled species reproduce so rapidly that the increased variation produced by sexual reproduction is not essential for adaptation to new conditions. Asexual species which reproduce more slowly may thrive so long as conditions remain fairly constant, but are less likely to survive sudden environmental changes. However (so Cherfas and Gribbon argue), this advantage is present only when the offspring of a single individual are very numerous and in intense competition with one another. Humans produce far too few offspring for the variety induced by sexual reproduction to have much effect upon the speed of evolution.[10] Besides, human survival has long since ceased to be primarily dependent upon the species' capacity to undergo biological change in response to new conditions. We survive environmental change by altering our technologies, not our genes. Thus, it would seem that for us asexual reproduction would be biologically more efficient, in that all parents would contribute to the gestation and nursing of the young.

In spite of their title, Cherfas and Gribbon do not believe that the human male is altogether biologically redundant. They suggest that the primary biological advantage of sexual reproduction in humans may be that the mixing of the genes of different individuals facilitates the alteration in each generation of the biochemical "passwords" whereby the body's defense systems distinguish between its own cells and those of foreign organisms. Without such continual change, disease organisms might successfully mimic those passwords and invade our bodies with impunity. Thus, asexual reproduction via cloning or parthenogenesis might eventually result in the extinction of the species through disease.

If this theory is correct, or if the mixing of genes from different individuals is biologically necessary for other reasons, then it is probable that neither cloning nor parthenogenesis can provide a satisfactory replacement for sexual reproduction in the human species. But the theory fails to demonstrate that males are biologically necessary. Reproduction through ovular merging would preserve whatever biological advantages accrue from the mixing of the genes of different individuals. Each child would have two parents, both female, and each contributing half of its genetic endowment. Moreover, in an all-female world, the relative muscular weakness of women would not be a serious disadvantage. Women's muscular strength is quite sufficient for independent survival, once the threat of male violence is removed.

Feminist dreams of an all-female world have been expressed in utopian novels at least since the early years of this century. In 1915, Charlotte Perkins Gilman published *Herland*, a fictional account of an isolated nation of women.[11] This unisex nation came about as a result of a rebellion 2,000 years ago by a group of female slaves, who rose up and slew their masters. One of these women proved capable of asexual reproduction, and her descendants repopulated the territory with a race of parthenogenetic females.

Gilman was a Darwinian evolutionist, but unlike Darwin—who believed that men are more highly evolved with respect to the distinctively human qualities of reason and imagination[12]—she held that it is women who represent the human "race type," i.e., who most clearly embody distinctively human qualities.[13] In her view, the male is a modification of that basic human type, specialized for sexual reproduction. The aggressiveness of males and their instinctive urge to compete against other males for status and domination are aspects of their sexual specialization. The female, in contrast, is specialized not for sex but for motherhood and cooperative social existence. The evolution of human intelligence was not a function of male aggression, as Darwin thought, but of the augmented capacity of human mothers to care for their helpless young. Mother love was the first social bond and the origin of productive human industry.

Herland illustrated these ideas. Its citizens followed a religion of maternal pantheism, in which all life was viewed as a continuous cycle of motherhood. This global motherliness "dominated society, ... influenced every art and industry, ... absolutely protected all childhood, and gave to it the most perfect care and training."[14] There was no violence, no competition, no poverty, and no irrational destruction of nature. Population was maintained at the level judged to be ecologically optimal, and undesirable genetic traits were eliminated through the voluntary choice of women carrying such traits to remain childless. One aspect of Gilman's utopia which most contemporary feminists would find implausible is the absence of genital eroticism. The women of Herland were not sexually attracted to one another (or to men, when these appeared), and had no desire for erotic experience.

Donna Young's *Retreat: As It Was!* presents a more modern feminist utopia, which is not lacking in eroticism.[15] Retreat is a planet colonized by an originally all-female race. Although they possess the technology for space travel, they favor less complex technologies in their daily lives, preferring the use of pack animals to motorized vehicles, and herbal medicines and psychic healing to what we might regard as more sophisticated forms of medical treatment. Most women have lovers, but reproduction is normally via parthenogenesis; once in several generations genes are shared in order to keep the genetic pool fluid. There is very little competition. War and internal violence are avoided, not just through the natural gentleness of women but also through training in nonviolence. As in *Herland*, the primary focus of the culture is the nurturing of children. This task is shared by all women rather than left to individual mothers. The downfall of this idyllic society is presaged by the appearance of the first male children as a result of a radiation-induced mutation. (For the idea is still current among some feminists that the male is an abnormal deviation from the basic human type.)

The Female Man, a novel by Joanna Russ, depicts a rather different female utopia.[16] Whileaway is a possible future world in which men are extinct and reproduction occurs through ovular merging. There is no warfare, but there is plenty of competition, and personal conflicts sometimes lead to deadly duels. Children are reared until the age of four or five in families of twenty to thirty women. After that, they leave home permanently for about six years of schooling and as many of wandering freely in search of adventure, before beginning to train for a profession. A lusty, nonmonogamous eroticism prevails.

The differences between these feminist utopias reflect different ideas about the nature of women, e.g., their eroticism and propensity for competition and violence. But in spite of their differences, all of these utopias have certain features in common. Regardless of the frequency of individual conflict, wars do not occur. Social relationships are egalitarian and nonhierarchical. Motherhood has high social status—in contrast to most patriarchal cultures, where it is treated as a liability which disqualifies most women from playing responsible roles in public life. Maternity and the values associated with it are not isolated within private families, but constitute the guiding ethos of the entire society. The natural world is not seen as something to be conquered, but as a mutually interdependent community of which we are part. Freedom from rape and random violence allows women to move as they choose through a world which is their own:

> There is no being *out too late* ... or *up too early* or *in the wrong part of town* or *unescorted*. You cannot fall out of the kinship web and become sexual prey for strangers, because there is no prey and there are no strangers—the web is world-wide.[17]

Perhaps an all-female society would have these utopian qualities and perhaps not. This would depend not only upon whether there really are certain innate psychological differences between the sexes, but also upon just how that society came into being and the conditions of its evolution. Suzy McKee Charnas's *Motherlines* depicts two very different all-female societies, both existing alongside a hyperpatriarchal society which enslaves and dehumanizes women.[18] One is a society of escaped female slaves, or "fems." The psychology of the fems has been shaped by their past enslavement, and they cannot reproduce without males. Thus, they never create a distinctive culture of their own. Although nominally free, they tend to replicate the master/slave relationships of the patriarchy. The Riding Women, on the other hand, are an independent people who reproduce parthenogenetically and have lived without men for centuries. Theirs is a nomadic culture, based on the use and worship of horses. They hunt for food and wage nonlethal ceremonial warfare, which serves more to unite than to divide the different tribes. Their society exhibits the egalitarian and communal values common to other feminist utopias.

A separatist female society might initially follow either of these psychological patterns. However, if the assumptions about female psychology which are shared by most feminist separatists are correct, it would tend towards the utopian pattern to the extent that it was able to free itself of masculinist influences. Unfortunately, we do not know whether those assumptions are correct. We do not know whether women are, on the average, naturally more nurturant and empathic or less belligerent and power-hungry than men. (In Chapter 5, we will take a closer look at the evidence which has been taken to show that males are naturally more aggressive.) We do know, however, that the psychological differences—innate or otherwise—*between* women are at least as great as the differences between the average woman and the average man. Thus, it is possible that an all-female society would recreate the hierarchical social structures typical of patriarchy, with the more dominant and aggressive women playing the "masculine" parts.

But it is pointless to pursue this debate about whether an all-female world would follow the feminist utopian pattern. For the total elimination of men is as improbable as the total elimination of women. It may be slightly more feasible in biological and economic terms, insofar as asexual reproduction by women may require less expensive technological facilities than asexual reproduction by men. But in social and psychological terms it is little more than a wistful fantasy.

The elimination of men, like the elimination of women, could conceivably occur through the effects of a sexually selective plague, loosed upon the world by mad scientists, or by the spontaneous emergence of some new virus. It could also conceivably occur through the mandate of a global dictatorship. But it is not going to come about through the voluntary decision of women. The likelihood that all women will ever agree to bear only daughters is vanishingly small. To be sure, women have better reasons for wanting to eliminate men than men have for wanting to eliminate women. Throughout history, women have been enslaved, abused and sometimes killed because of their sex, and the end of sexism and gendercide is nowhere in sight. Yet very few women see (all) men as the enemy. Most women realize that although men may profit from patriarchy and maintain it through their daily actions, most men are not evil beings. Some men are feminists, just as some women are antifeminists. Most feminists do not want to eliminate men, but to build a world in which women and men can live as equals.

Biological Androgyny

A third radical alternative is the elimination of both sexes in favor of some sexually indeterminate or intermediate human form or forms. Male physiological development is normally triggered by information carried on the Y-chromosome, which causes testicular development in the male fetus. The testicles then secrete testosterone, which leads to other developments in the direction of biological masculinity. However, all human fetuses, whether they have a Y-chromosome or not, will develop an essentially female physiology unless they are exposed to sufficient quantities of the hormone testosterone at a critical prenatal stage.[19] (This stage is believed to be about six weeks gestational age.) Conversely, any normal human fetus, whatever its chromosomal sex, will develop a basically male physiology if it is exposed to enough testosterone at this critical stage.[20] And any otherwise normal fetus which is exposed to more testosterone than is the normal female but less than the normal male will develop a genital and bodily morphology which is intermediate between the normal male and female forms.

Such sexually intermediate individuals are sometimes called bio-logical androgynes, or pseudo-hermaphrodites ("pseudo" because they are rarely fertile in both modes). Biological androgynes may be produced either by natural or by iatrogenic (i.e., medically induced) accident. The usual current medical practice is to "correct" such individuals by assigning them to one sex to the other, if necessary surgically altering the genitals to conform to that sex, and/or administering hormone therapy to encourage physical development in an unambiguously masculine or feminine direction. But medical science could be used to deliberately create human androgynes, were it not universally regarded as unethical to do so. The sex-selection technologies of the future could offer not just two options, but three or more. The production of a biologically androgynous human race, until now only a logical possibility, is rapidly becoming a technical possibility as well.

As humanity gains means of perpetuating itself without sexual reproduction, the distaste which most people feel towards biological androgyny might gradually dissipate, and its potential advantages might be recognized. As we have seen, both masculine and feminine physiologies entail certain biological disadvantages: e.g., a shortened life span in the former case and lesser upper-body strength in the latter. In time it might become possible to design and produce human beings who possess the physiological advantages of both sexes, while avoiding the disadvantages of each. There may also be mental or psychological disadvantages inherent in both male and female physiology. Some researchers believe that the typically greater female precocity in the development of verbal skills, and the typical male superiority in spatial skills are the result of inherent differences between the structure and organization of male and female brains.[21] A biological androgyne might possess the mental advantages characteristic of both sexes. Moreover, the universal adoption of physical androgyny might be one way of producing a world without sexual privilege.

It might seem obvious that biological androgyny is not necessary for the elimination of male domination. Psychological androgyny— i.e., the combination of stereotypically "feminine" and "masculine" psychological traits in each individual, female or male—might be sufficient. Contemporary feminist androgynists do not advocate the elimination of biological gender, but rather the elimination of all those cultural factors which serve to mold women and men into contrasting psychological patterns. A world of psychological androgynes might achieve complete sexual equality without eliminating biological gender.[22]

But the ideal of universal psychological androgyny as a model for human liberation has many problems. One is that it is doubtful that psychological androgyny—even if it could be achieved by all persons—would automatically lead to the alteration of male-dominated social institutions to permit equal participation by women. Male domination does not persist primarily because women fail to develop their so-called "masculine" capacities (leadership, ambition, competitiveness, etc.). The reverse is much more likely to be true: as a subordinate caste, women have relatively little opportunity to develop such traits, and generally incur social disapproval if they do. To attempt to change power relationships by changing individual personalities is to put the psychological cart before the cultural horse. We must change the power relationships created by patriarchy before we can expect to eliminate the personality traits which reflect them.

Another problem is that complete psychological androgyny may be impossible so long as biological gender persists. Male and female brains may indeed be different. Moreover, the greater size and muscular strength of males may inevitably shape the psychology of women and men into somewhat different forms, at least so long as both sexes coexist within the same society. It is a plausible hypothesis that a person's size and upper-body strength tend to influence the ways in which s/he deals with other persons. For a small and relatively weak person, physical coercion or threat is less often an option than for a larger and stronger person. This fact may have a pervasive influence upon social interactions even in societies where violence is relatively rare, for the peace may be maintained by a pervasive pattern of deference. Lethal weapons such as knives and guns, which can be used even by small persons, do not negate the social importance of size and strength, since relatively few personal disputes are settled by the use of such weapons.

Even if biology makes no direct contribution to the psychological differences between the sexes, the innumerable gender-differentiating elements present in all cultures may prove even more difficult to eliminate than biological gender itself. It is no longer obviously true that human cultures can be altered more easily than human biology. Biology can sometimes be altered simply by taking a pill; but cultural forms may be defended to the death, in spite of powerful moral and pragmatic arguments in favor of change. Thus, it is worth con-

sidering the option of biological androgyny, even if in the end it must
be rejected.

In *The Left Hand of Darkness*, Ursula LeGuin presents a fictional
account of what a society of human androgynes might look like.[23]
Winter is a cold planet which has been colonized by people who, most
of the time, are neither male nor female. Once a month, each person
goes into "kemmer," a phase in which s/he becomes sexually aroused
and fertile in either the male or the female mode. When lovers enter
kemmer at the same time, one becomes male and the other female, but
neither can know in advance which s/he will become. Thus, anyone
may either sire a child or bear one, and many do both, at different
times. The resulting society has many of the features which are typical
of feminist utopias. Individuals and groups compete for honor and
prestige, and murders sometimes occur, but there are no wars. The
population is small and stable. And, of course, there is no sexual
inequality:

> Anyone can turn his hand to anything. This sounds very simple, but its
> psychological effects are incalculable. The fact that anyone between
> seventeen and thirty five or so is liable to be ... "tied down to childbear-
> ing," implies that no one is quite so thoroughly "tied down" here as
> women, elsewhere, are likely to be—psychologically or physically. Burden
> and privilege are shared out pretty equally; everybody has the same risk to
> run or choice to make. Therefore nobody here is quite as free as a free male
> anywhere else ...
>
> There is no unconsenting sex, no rape. As with most mammals other
> than man, coitus can be performed only by mutual invitation and
> consent ...
>
> There is no division of humanity into strong and weak halves,
> protective/protected, dominant/submissive, owner/chattel, active/pass-
> ive. In fact the whole tendency to dualism that pervades human thinking
> may be found to be lessened, or changed, on Winter.[24]

The vision of an androgynous world has inspirational value.
Androgyny represents the fulfillment of an ancient human dream: the
dream of bridging the chasm which separates male and female
existence, confining each of us to just half of the human experience.
Androgynous deities have been prominent in many religions. In some
religious traditions (e.g., that of the Sioux), priests or shamans adopt
the attire of opposite sex, becoming, in effect, an intermediate
gender.[25] Androgynous styles of clothing, coiffure and make-up are
currently popular in many western cities. Unlike most of the unisex

styles of earlier decades, the new androgynous looks often effectively disguise the individual's biological sex. The impression conveyed is one of gender fluidity, unconfined to either of the traditional options.

In a world of biological androgynes, individual personalities would still differ, and feminine and masculine roles might not entirely disappear. But they would inevitably become more optional and more readily reversible. No one would be forced to repress one half of her human potential simply because she happened to be born with a certain genital configuration. Certain personality traits might continue to be stereotyped as feminine or masculine, but it would be impossible to avoid the recognition that these are all essentially human traits, potentially present in each individual. Biological families might still constitute the basic social unit: life partnerships might be formed and partners might choose to have children and to rear them in a private domestic context. But one parent would no longer automatically have a prima facie power advantage over the other because of their different genders, and the family would no longer function as a training ground in sexual inequality.

But the creation of a biologically androgynous human world is even more improbable than the creation of an all-male or all-female world. The ethical obstacles alone are apt to prove insuperable. The unforeseen side effects of estrogen replacement therapy, birth control pills and other medical uses of sex hormones have taught us that it is dangerous to tamper with the normal hormone balance. There is every reason to fear that the radical hormonal or genetic manipulations necessary to produce biological androgynes might have harmful effects upon human health. The research necessary to find out whether or not this is true would require experimentation on human subjects too young to give informed consent. Research on nonhuman animals would be insufficient to demonstrate the safety of any particular procedure, since every species is somewhat different from every other. The experimental imposition of serious risks upon human beings who are incapable of consenting to those risks may be justified when those individuals are already suffering from severe illness, and when the experimental procedure is more likely to improve their condition than to worsen it. But it is impossible to justify experimental procedures which may be harmful to actual or future persons, and which have no potential therapeutic value.

Even if it were possible to genetically engineer healthy human androgynes without any violation of human rights, other factors would almost certainly prevent the conversion of future generations to universal biological androgyny. If androgynes proved incapable of uterine gestation, the excessive cost of artifical gestation would probably necessitate the retention of women as reproducers. In any case, ideological inertia would be likely to prevent the conversion of humanity to that model. Rational argument may never suffice to persuade the majority of persons that biological gender is a contingent rather than an essential feature of personhood, or that we would be better off without it. Whatever the inherent advantages of biological androgyny, parents might wisely hesitate to produce children who will be perceived by many people as subhuman monsters.

Thus, a universal conversion to biological androgyny—like a universal conversion to all-male or all-female offspring—could probably be brought about only by coercive means. The harmful consequences of such coercive means would outweigh any possible benefits. Besides, there is no evident reason why any global authority would want to bring about any of these radical options, given the cost involved and the inevitable resistance. Tyrants rarely object to sexual inequality, because it serves their needs well, preparing people to accept other institutions based on dominance and submission. Thus this option, like the previous ones, is likely to remain in the realm of science fiction indefinitely.

Unisex Communities

Unlike the other extreme options, the formation of self-reproducing unisex communities is bound to occur. Such communities will differ from the male monasteries and female nunneries which exist in many religious orders, in that their members will be able, without forming sexual liasons with members of the opposite sex, to have children which are genetically theirs and guaranteed to be of the same sex. This is already possible on the basis of present reproductive technologies. Indeed, such communities may already exist on a small scale. If so, it would be understandable that they should shun publicity.

If self-reproducing all-female communities do not already exist, they are likely to appear very soon. Such communities might emerge through the formation of alliances between unisex families. Many female couples are already rearing children, although it is not yet possible for them to have children who are each genetically parented by both partners. Women who wish to avoid sexual relationships with men can become pregnant through artificial insemination. Many physicians refuse to provide that service to unmarried women. However, sperm can sometimes be obtained from feminist-run clinics, or male friends. Women who do not wish to have sons can sometimes obtain prenatal diagnosis of fetal sex, and abort the fetus if it is male. The development of techniques of parthenogenesis or ovular merging would greatly encourage the formation of all-female families, since offspring produced by these methods would necessarily be female. Given the biases of the "straight" community, it would be natural for such unisex families to combine forces, sharing housing and child care, and perhaps operating farms or other businesses to obtain economic self-sufficiency. Some might choose to educate their own children in order that they might be relatively free of peer-group abuse and heterosexist indoctrination.

Although some male couples will choose to have and rear children, the deliberate formation of all-male families or larger self-reproducing all-male communities will probably be less common than the deliberate formation of all-female families and communities. The formation of all-male families presents a more difficult problem. Men who can obtain the services of a surrogate mother can have children of their own begetting without entering into a sexual relationship with a woman. In theory, they could even specify that the fetus must be aborted if female, or that they will adopt it only if it is male. But the legality of such a contract would be highly questionable—as is surrogate motherhood itself, in many jurisdictions. Ectogenesis, and more reliable methods of preconceptive sex selection could alleviate these problems. However, relatively few men are eager to undertake the primary responsibility for child rearing, unless they already have children and circumstances leave them no other option. This could change, if child care ceases to be seen as unmanly occupation; but at present most men seem content to leave the rearing of children to women.

Thus, the group most likely to form self-reproducing unisex communities are the feminist or lesbian separatists. Many feminist separatists are "lesbian" in the strictly sexual sense of the word. But most of those who define themselves as lesbians do so not simply because they prefer women as lovers, but because of their personal and political identification with the interests of women as an oppressed group.

Women have always had more or less distinctive "cultures," which have been overshadowed but never entirely suppressed by dominant male cultures. Women's attitudes towards life and death, nature, children, politics, and personal relationships tend to be somewhat different from those of men. Carol Gilligan has explored some of these differences.[26] Gilligan argues that women's moral development and ways of conceptualizing moral problems tend to differ from those of men—which are usually taken as the norm. Men tend to conceive of morality as a structure of rules and principles whereby competing interests and claims can be placed in rank order, and the less valid ones dismissed. Women are more apt to acknowledge competing claims and responsibilities, striving to resolve conflicts in ways which avoid harming any individual. Their morality, Gilligan argues, is based more on care and connection than on abstract conceptions of justice, or adherence to rules and principles.

Spiritual feminists believe that not only is women's approach to moral issues distinctive, but so is their "spirituality," or mode of relationship to reality as a whole. Carol Ochs argues that women's experiences provide insights into the life process that have been omitted from most religions, which take account only of men's experiences.[27] Her hypotheses, like Gilligan's, are consistent with the ideas which have been dramatized in the various all-female utopias. In Ochs's view, women's experiences of pregnancy, birth, mothering caring for the sick and dying, and living as outsiders in male-dominated societies provide an ideal context for the development of a spirituality which is oriented towards life in this world, rather than towards preparation for some future life. For those whose spirituality is anchored in female experiences, "salvation" is found through the daily process of living with and caring for other persons, not through a solitary journey towards some transcendant goal. Like the pantheist philosopher Baruch Spinoza, Ochs defines God not as a being which is distinct from the world, but as the sum total of reality—its material and biological as well as its mental aspects.[28]

To the extent that women do tend to have a distinctive culture, separatist women's communities may make a vital contribution not just to the women's movement but to the future of life on this planet. Sex selection is not essential for the establishment of all-female communities. However, the option of bearing and rearing only daughters might facilitate the long-term survival of such communities. Transgenerational separatist communities would provide a more significant test of the hypotheses about women's distinctive culture— or their potential for the creation of such a culture—than more transitory separatist communities can. In such communities, women's cultures could evolve more freely than they can among women who are dependent upon men for economic and emotional support, employment, physical protection, spiritual guidance and sexual satisfaction.

The existence of stable separatist communities would be valuable even for women who have no desire to join them—just as the existence of the state of Israel is a source of pride and self-respect even for those Jews who live elsewhere. For the first time, women would have a practical alternative to living as "others" within patriarchal society. Eventually, some women's communities might even become large enough successfully to demand complete or partial political independence.

The establishment of separatist women's communities will not be easy, or their success guaranteed. Such communities may suffer from many of the problems which have plagued socialist and religious communities established within larger societies. Lacking legally constituted police powers to compel conformity to majority decisions, they may be vulnerable to schisms and internal rebellion. Hostility or aggression from outside may lead to a defensive, introverted posture. Not all the children born within the community will elect to remain within it, and adult converts may have difficulty fully accepting its norms and mores. Utopian separatist societies will have to evolve gradually through often-painful processes of trial and error. But it is well to remember that one of the chief problems which have contributed to the failure of so many other attempts to establish stable utopian communities has been the retention of male-supremacist customs and attitudes, which has prevented the development of genuinely egalitarian social forms. In this respect, at least, all-female communities might succeed where other experimental societies have failed.

Returning to Reality

Having glanced at some of the extreme options which might be pursued via sex selection and other new reproductive technologies, we must now return to the world as it is and to the social realities which women and men will continue to experience for the foreseeable future. The future is less foreseeable than it has ever been, and the new reproductive technologies contribute to its uncertainty. But it is certain that at least for the next several generations most people—assuming that there will be people on earth at all—will continue to be born into sexually mixed families and communities. Moreover, patriarchy and male domination will continue to be a salient feature of most families and communities, even if the feminist movement remains strong. It is in this realistic context that we must consider the objections to sex selection.

Notes

1. Thomas Aquinas, *Summa Theologica* (translated by fathers of the English Dominican Province, New York: Benzinger Brothers, 1947), Volume I, 447.

2. See, for instance, Nancy Chodorow, *The Reproduction of Mothering: Psychoanalysis and the Sociology of Gender* (Berkeley: University of California Press, 1978), and Dorothy Dinnerstein, *The Mermaid and The Minotaur: Sexual Arrangements and Human Malaise* (New York: Harper and Row, 1976). Chodorow and Dinnerstein maintain that the fact that children are reared by women is the cause of many of the typical personality differences between the sexes, and that this fact also explains why both men and women tend to be uncomfortable about women in positions of power.

3. Simone de Beauvoir, *The Second Sex*, (translated by H. M. Parshley (New York: Random House, 1974), 38.

4. See Jo Durden-Smith and Diane de Simone, *Sex and the Brain* (London: Pan Books, 1983), 113.

5. Philip Wylie, *The Disappearance* (St. Albans: Panther Books, 1974).

6. Frank Herbert, *The White Plague* (London: New English Library, 1983), 358.

7. de Beauvoir, 160. (Of course, it is not *quite* true that no man would consent to be a woman. Some men do submit to sex change operations, but these are usually men who believe themselves to be women already, i.e., psychological females trapped in male bodies.)

8. Mia Albright, "Centralizing Feminism: Where Is the Power Center of the World's Populations of Womyn? A Nationalist Feminist Perspective," *Fourth Women and Labour Conference Papers, Brisbane 1984* (St. Lucia, Queensland: The Organizing Committee, Fourth Women and Labour Conference, 1984), 3.

9. Jeremy Cherfas and John Gribbon, *The Redundant Male* (London: The Bodley Head, 1984), 179.

10. Cherfas and Gribbon, 38.

11. Charlotte Perkins Gilman, *Herland* (London: The Woman's Press, 1979).

12. Charles Darwin, *The Descent of Man and Selection in Relation to Sex* (New York: D. Appleton, 1874), 577.

13.   Charlotte Perkins Gilman, *His Religion and Hers: A Study of the Faith of Our Fathers and the Work of Our Mothers* (Westport, Connecticut: Hyperion Press, 1976), 82.

14.   Gilman, *Herland*, 73.

15.   Donna J. Young, *Retreat: As It Was!* (Weatherby Lake, Missouri: The Naiad Press, 1979).

16.   Joanna Russ, *The Female Man* (London: W. H. Allen and Company, 1979).

17.   Russ, 81.

18.   Suzy McKee Charnas, *Motherlines* (New York: Berkeley Publishing Corporation, 1979).

19.   It appears, however, that a second X-chromosome is necessary for female fertility. Individuals born with Turner's Syndrome, who have only one X-chromosome and no Y-chromosome, develop as infertile females.

20.   Some individuals are insensitive to the usual effects of testosterone, and thus will develop a feminine physiology regardless of how much of that hormone they may be exposed to.

21.   See, for instance, Durden-Smith and de Simone, op. cit.

22.   See *"Femininity,"* *"Masculinity,"* and *"Androgyny,"* edited by Mary Vetterling-Braggin (Totowa, New Jersey: Littlefield, Adams and Company, 1982) for further discussion of the feminist concept of psychological androgyny.

23.   Ursula K. LeGuin, *The Left Hand of Darkness* (New York: Ace Books, 1969).

24.   LeGuin, 93–94.

25.   See Elemire Zolla, *The Androgyne: Fusion of the Sexes* (London: Thames and Hudson, 1981), 82–83.

26.   Carol Gilligan, *In A Different Voice: Psychological Theory and Women's Development* (Cambridge: Harvard University Press, 1982).

27.   Carol Ochs, *Women and Spirituality* (Totowa, New Jersey: Rowman & Allanheld, 1983).

28.   Ochs, 95.

4

Nonconsequentialist Objections to Sex Selection

The objections considered in this chapter do not involve any particular predictions about the social or psychological consequences of sex selection. Instead they focus upon the *act* of preselecting a child's sex, and upon certain features which that act may be thought to have, and which may be thought to be morally objectionable. Any action may be morally wrong under certain conditions. The question is whether any of these objections to sex selection show that it is morally objectionable in itself, and thus always wrong in the absence of some unusual extenuating circumstance.

Because preconceptive sex selection and selective abortion are rather different acts, it is best to consider the objections to selective abortion separately. We will begin by examining three objections which are equally applicable to either type of sex selection and then turn to the objections to abortion, both in itself and as a method of sex selection.

Is Sex Selection Unnatural?

Often, the first reaction to the idea of preselecting a child's sex is that this should not be done *because it is unnatural*. But what does it mean to say that a human action is unnatural? And why should we assume that actions which are unnatural are for that reason immoral? The argument that sex selection is wrong because it is unnatural can succeed only if there is an interpretation of the term "unnatural" which enables us both to clearly distinguish between natural and unnatural actions, and to understand what there is about the latter which is morally objectionable. It is easy to produce an interpretation which meets one of these requirements, but I know of no interpretation which meets both. On some interpretations it is easy to tell what is natural and what is unnatural, but there is no evident reason why we ought to object to what is unnatural. On other interpretations, it is clear that we should avoid doing what is unnatural, but there is no simple way to decide which actions are unnatural.[1]

On one interpretation, an action is unnatural if it would not occur in a "state of nature," i.e., prior to the invention of sophisticated technologies and forms of social organization. It is clear that prenatal and preconceptive sex selection are unnatural in this sense, since the necessary technologies would not exist in a state of nature. But neither would any of the rest of our post-Paleolithic science and technology. We cannot object to sex selection on this basis unless we wish to advocate a return to a simple gathering-and-hunting lifestyle. This is obviously not feasible for the great majority of human societies, and probably never will be unless some disaster reduces the population to a small fraction of its present level.

Another common meaning of "unnatural" is "contrary to long-established habit or custom." This is the sense in which one might say that wearing lipstick is unnatural for most men. On this interpretation, too, the new methods of sex selection are clearly unnatural. They cannot have been customarily used in the past, since they did not exist in the past. (Of course, postnatal sex selection—by means of female infanticide or the neglect of female children—has been customary in a great many societies.) But the fact that a new technology enables us to do something which we could not do before is hardly a persuasive moral objection to the use of that technology. Thus, if "unnatural" is taken to mean only "uncustomary," no valid objection to sex selection can be derived from the fact that it is unnatural.

A third common meaning of "unnatural" is abnormal, i.e., indicative of individual or social pathology. This is the meaning which those who claim (implausibly) that homosexual behavior is unnatural seem to have in mind. On this interpretation, what is unnatural is by definition undesirable. If sex selection is pathological, then of course we should avoid it. But how can we decide whether or not sex selection is indicative of pathology? It is implausible to hold that the desire to preselect the sex of one's children is always a sign of *individual* psychological pathology. For the social and economic realities of patriarchy often create conditions in which son- or daughter-preference is rational from the viewpoint of the individual.

The suggestion that sex selection is indicative of *social* pathology may appear to be more plausible. Son-preference is largely the result of sexist attitudes and institutions. A sexually egalitarian society would probably have little use for methods of sex selection, except perhaps to prevent the birth of children with sex-linked genetic illnesses. But it does not follow from this that sex selection is immoral. For, depending on the ways in which it is used, it may be part of the problem or part of the solution—or both. It might, for instance, be used by feminists to produce daughters, in order to offset its use by other persons to produce males. In this case, it would still be symptomatic of past and present injustices, but this would not show that it is morally objectionable.

A fourth interpretation of the claim that sex selection is unnatural is that it is in some way contrary to innate human needs. Extremely tight corsets and overly crowded living conditions are other things which might be regarded as unnatural in this sense. Human beings, it might be argued, have an innate need to breathe freely and to have some opportunity for privacy. There is, however, no reason to believe that human beings have an innate need not to select their children's sex. The history of female infanticide surely tells against that suggestion. It is possible that people of future generations will be harmed by the use of sex selection. But, if so, they will probably be harmed not because sex selection is inherently contrary to some basic human need, but rather because it has been unwisely used.

There are many other possible interpretations of "unnatural," but none which can make this argument against sex selection valid. On some interpretations the claim that sex selection is unnatural is true but morally irrelevant. On other interpretations, it is clearly false. The distinction between natural and unnatural behavior is not only very unclear, but of very little relevance to moral issues. We are certainly a part of nature, but we are not a species whose behavior is narrowly determined by our biological inheritance. Simone de Beauvoir is half right when she says that,

> Humanity is not an animal species, it is a historical reality. Human society is an antiphysis—in a sense it is against nature; it does not passively submit to the presence of nature but rather takes over the control of nature on its own behalf.[2]

The truth is that humanity *is* an animal species, but one whose nature includes a highly creative intelligence. It is the opposite of the truth to suppose that the development and use of new technologies, including new methods of sex selection, are contrary to human nature. It *is* human nature to invent ways of doing things we were not able to do in the past.

Is Sex Selection "Playing God"?

Another common reaction to sex selection is that to preselect our children's sex is to "play God"—an activity which is presumably not to be condoned. This objection, like the objection that it is unnatural, most often expresses an intuitive reaction rather than a clearly reasoned moral position. What might it mean to say that sex selection is playing God?

On a literal interpretation, this claim means that human beings have no right to select the sex of their children, because there is some supernatural being which claims that right for itself, and forbids us to trespass upon its prerogatives. The problem with this argument is that there is no reason to believe that its premise is true. Even if there should turn out to be valid evidence for the existence of a god of the sort envisioned, it will still be extremely difficult to prove that this god demands the exclusive right to determine the sex of human children. Furthermore, it will remain an open question whether or not that god has the moral right to such a monopoly. Unless might makes right, we are entitled to ask why any external authority, human or divine, should be accorded the exclusive right to control the sex of our children, without our individual or collective consent.

But it is not necessary to interpret this objection in such a literal way. When someone responds to an obscure question by saying, "God only knows!" they usually mean only that they do not know the answer, and that in all probability no one else does either. Similarly, the "playing God" objection may be an elliptical way of saying that no one has the right to select a child's sex: that the matter should be left to chance, fate, or whatever gods may be. On this interpretation, however, the objection amounts to little more than an expression of disapproval; no specific grounds for that disapproval have been provided. Perhaps the assumption is that human beings are not wise enough to use such a power responsibly. But that assumption will have to be proved by further argument.

I suspect that the "playing God" objection gains much of its apparent force from assumptions about the possible consequences of sex selection. On this interpretation, sex selection is implicitly compared to other unprecedented alterations of the human or natural environment, such as changing the weather—alterations whose long-term consequences we cannot know in advance, but which may be detrimental.

There is, however, another interpretation of this objection, which is not reducible to an appeal to its possible consequences. The desire to preselect a child's sex may be thought to reflect a morally objectionable attitude towards the potential child. The President's Commission for the Study of Ethical Problems in Medicine and Biomedical and Behavioral Research argues that sex selection involves an attitude which: "taken to an extreme, ... treats a child as an artifact and the reproductive process as a chance to design and produce human beings according to parental standards of excellence."[3]

Although this objection is made in the context of sex-selective abortion rather than preconceptive sex selection, it would seem to be equally applicable to either. The Commissioners have no moral objection to the use of genetic screening or other medical interventions "to correct or avoid unambiguous disabilities."[4] Such interventions are readily justifiable by the improved health and well-being of children. It is clearly better that a normal child be born than one who will be badly handicapped; and sometimes it is better that no child be born than one who will be severely abnormal. In their view, however, to go beyond the prevention of abnormalities to the implementation of preferences for this or that type of (normal) child is unacceptable, because it treats the child as if it were an object to be shaped according to the personal whims of others.

Implicit in this objection is the Kantian principle that persons must always be treated as ends in themselves and never merely as means. If sex selection necessarily involves treating a future person as a mere means to someone else's ends, then it must be regarded as immoral; or so a Kantian ethicist might argue.

What are we to make of this argument? It is certainly true that the desire to preselect a child's sex might be a symptom of a morally objectionable attitude towards the child. A prospective parent who has a strong preference about the sex of a future child may be regarding the child as raw material to be shaped into a specific mold, regardless of the child's own needs or preferences. For instance, a man may be set on having not just a son, but a son who will look, think, and act just like his father. If so, he may fail to respect the child's own individuality, regarding the child as little more than an extension of himself.

But is the desire to preselect a child's sex inevitably associated with such a selfish attitude towards the child? I see no reason to believe that it is. There is a clear distinction between wishing to have a child of a particular sex and regarding the child as something less than a potentially autonomous person. Prospective parents who make use of sex selection methods may be quite well aware that the child will be a unique and autonomous individual, not a mere thing to be shaped according to their subjective preferences.

There is, indeed, no a priori reason why the decision to preselect a child's sex might not be motivated by an unselfish desire to ensure that the child will have the best possible life. Suppose that a couple choose to have a son because they have a deceased relative who has willed his estate to go only to a son of theirs, and only after their deaths, and they want their child to benefit from that inheritance. That they reason in this way is hardly evidence that they are viewing the child as nothing more than a means for achieving their own selfish ends. Granted, they might be deceiving themselves about the real reasons for their preference for a son; they might really be motivated by some advantages which they foresee for themselves. But there is no reason to insist that this must be so.

Perhaps there is yet another way of interpreting the claim that sex selection is playing God, which makes this claim into a valid moral objection. However, none of these interpretations does. The "playing God" objection, like the objection that sex selection is unnatural, appears to be based upon little more than a badly reasoned fear of novelty.

Is Sex Selection Inherently Sexist?

Perhaps the most troubling of the nonconsequentialist objections to sex selection is that it is inherently sexist. Sexism is usually defined as wrongful discrimination on the basis of sex. Discrimination based on sex may be wrong either because it is based on false and invidious beliefs about persons of one sex or the other, or because it unjustly harms those discriminated against. The nonconsequentialist objection to sex selection is that it is a sexist act because it is invariably motivated by sexist beliefs.

To some, it appears self-evident that only a person who is in the grip of sexist ideology would prefer to have a son rather than a daughter, or vice versa. Tabitha Powledge bases her case against the development and use of new sex selection methods on this point. She argues that,

> we should not choose the sexes of our children because to do so is one of the most stupendously sexist acts in which it is possible to engage. It is the original sexist sin. ... To destroy an extant fetus for this reason is more morally opprobrious than techniques aimed at conceiving a child of a particular sex, but they are both deeply wrong. They are wrong because they make the most basic judgment about the worth of a human being rest first and foremost on its sex. [For this reason] we should not put public money into the development of pre-conceptive sex-choice technology .. and .. such technology, if it does become available, should not be used.[5]

In this form, the argument is unsound. It is false that all persons who would like to preselect the sex of their children are motivated by the belief that members of one sex are more valuable. Some people, for instance, would like to have a son because they already have one or more daughters (or vice versa), and they would like to have at least one child of each sex. Others may believe that, because of their own personal background or circumstances, they would be better parents to a child of one sex than the other. On the surface, at least, such persons need not be motivated by any sexist beliefs.

However, it may be argued that the desire to preselect sex is always based on covertly sexist beliefs. Michael Bayles notes that the desire for a child of a particular sex is often instrumental to the fulfillment of other desires, such as the desire that the family name be carried on. Such instrumental reasons for sex preference, he argues, are always ultimately based upon irrational and sexist beliefs. In many jurisdictions, it is no longer true that only a man can pass his family name to his children. Thus, in those jurisdictions it would be irrational to prefer a son for this reason. Even the desire to have a child of each sex is irrational, according to Bayles, because there are no valid reasons for supposing that this would be better than having several children of the same sex. He considers the case of a man who already has two daughters and would like to have a son as well, "so that he could have certain pleasures in child-rearing—such as fishing and playing ball with him."[6] This man would be making a sexist assumption, since he could perfectly well enjoy such activities with his daughters.

John Fletcher also argues that the desire to preselect a child's sex (except for certain medical reasons) can only be based upon irrational and sexist beliefs. He says that,

> Prima facie examination of any argument for sex selection cannot overcome the unfair and sexist bias of a choice to select the sex of a child. The desire to control the sex of a child is not rational, since any claim that is made for the parents' preference for one sex can be demonstrated to be provided also by the other sex.[7]

But is it really true that there are no nonsexist reasons for preselecting sex, other than those medical reasons which are not in question here? Consider the situation of women in harshly patriarchal societies, such as that of northern India. Horowitz and Kishwar interviewed women in a rural village in Punjab. Many of them said that while they would like to have daughters for companionship and help with domestic work, they are reluctant to bring a female child into a society in which she will be abused and devalued, as they themselves have been.[8] It is implausible to argue that their preference for sons is irrational, or necessarily an indication of sexism on their part. Their motivations are at least partly altruistic, and do not appear to be based on any false beliefs.

A less altruistic reason for son preference in northern India is that a son is an economic asset, while a daughter usually is not. A daughter will almost certainly earn far less than a son, if she enters the job market at all. And, of course, there will be the expense of providing her with a dowry. It would surely be wrong to blame a couple who decide not to have children because they cannot afford to raise them. Why, then, should we blame them because they decide not to have daughters, for the same reason?

If son-preference is rational in rural north India, and not necessarily a sign of sexism on the part of the prospective parents, then it will be difficult to argue that this is not also true in much of the rest of world. Wherever son-preference is especially pronounced, it is due in large part to powerful economic motivations. Even in societies which provide some social support for the aged, sons are often an important part of old-age security.

It will not do to argue, as Fletcher does, that such economic motivations for son-preference are irrational because "Few jobs exist that women cannot perform as well or better than men when performance is the criterion of evaluation."[9] While this is certainly true, the fact remains that women's average earning capacity is far from commensurate with the true value of their work. The average full-time employed woman in America earns just 59 percent as much as the average man, and the average women with a college degree still earns less than the average man with only a high school education. Poor women, especially if they have children, have few opportunities of escaping poverty. The morally appropriate social response to this situation is to remove such economic incentives for son-preference, e.g., by eliminating unjust discrimination against women in education, hiring and promotion, and providing more adequate unemployment, old-age, and disability support for all persons. But, where such changes have not yet occurred, it might be argued that poor people have the right to seek to better their economic status by having sons rather than daughters.

In response to this point, it might be said that by adopting this strategy poor parents are attempting to exploit the sexism of the society for their own economic gain. Letty Pogrebin, after listing a dozen reasons which are sometimes given for son-preference, concludes that there are just two real reasons. One of these is that "parents want to cash in on patriarchal privilege."[10] The implication is that to do this is morally objectionable. It is, no doubt, often immoral to seek to profit from an unjust situation. But is it *always* wrong to do so? If poor parents can benefit themselves and their families by selecting sons, and if this does no direct harm to anyone else, then it is not clear that what they do is in any way unjust. If their actions are morally objectionable it can only be because of their unintended social consequences.

There are many other nonsexist reasons for son- or daughter-preference. Even in the industrialized nations, prospective parents may have reasons to prefer that their children, for their own sake, be male. Women are still far from enjoying the full range of freedoms and opportunities available to men. On the average, they not only earn much less but work much longer hours, because regardless of whether they have jobs they are still expected, in most cases, to shoulder much heavier domestic responsibilities. Male violence still makes the lesser size and upper-body strength of women a serious liability. The threat of rape still limits women's freedom of movement. So long as these many forms of oppression persist, it is absurd to suggest that women are guilty of sexism if they wish to have male children in order that the latter may enjoy the freedoms which women are still denied.

I am not suggesting that most women reason in this way, still less that most women ought to. Many prospective mothers would be equally content with a child of either sex; and many others would prefer a daughter. Of these, some are planning to raise a child without a male partner and believe that under the present conditions they would have more in common with a child of their own sex and thus (they hope) a better relationship with her. A son might be able to share most of their particular interests and activities, but he could not share the basic experience of being female in a society which still values males more highly. However much he may sympathize with the plight of women, he will still be a member of the more privileged sex. While such expectations are bound to prove mistaken in some cases, I see no grounds for condemning them as either sexist or irrational.

Other women may prefer to have daughters because they fear that, in Sally Gearhart's words,

> if they have sons no amount of love and care and nonsexist training will save those sons from a culture where male violence is institutionalized and revered. These women are saying, "No more sons. We will not spend twenty years of our lives raising a potential rapist, a potential batterer, a potential Big Man."[11]

As a group, men are more likely to resort to serious violence than are women. Is it morally wrong to take account of such proven statistical differences between the sexes in deciding whether and how to make use of the new methods of sex selection? Most feminists would agree that it is usually unjust to discriminate against individuals of either sex on the basis of merely statistical differences between the sexes. Individuals have the right to be judged on their own merits, not condemned by association with some group to which they happen to belong. But choosing to have a daughter rather than a son on the grounds that females tend to be less violent is not an instance of such injustice. The son one might have had instead might or might not have turned out to be violent; but since he does not exist there is no way to evaluate him as an individual. Furthermore, since he does not exist he cannot have been treated unjustly. This is most clearly true in the case of preconceptive sex selection. We will presently consider the case of sex-selective abortion.

Thus, while many reasons for preselecting sex are sexist, many others are not. It follows that the use of sex selection is not an inherently sexist act. That some parents wish to preselect their children's sex is largely a result of sexism at the social level. But this does not show that what *they* do is wrong.

Is Abortion Immoral?

If abortion is itself immoral, then it follows a fortiori that sex-selective abortion is wrong. On the other hand, if abortion is something which women have a right to choose, it does not follow that sex-selective abortion is morally acceptable. For there are some things which people have a right to do but which they nevertheless ought not to do. Thus, the morality of abortion itself and the morality of sex-selective abortion are distinct questions.

I will begin by stating the case for regarding voluntary early abortion as an option which women have the right to choose for any reasons they consider valid. Next I will consider the moral distinction between early and late abortion. I will argue that while the latter does require somewhat stronger justifications, the choice of abortion at any stage of pregnancy is within a woman's moral rights. Finally, I will return to the question of whether sex-selective abortion is morally objectionable.

A central question in the debate over the morality of abortion is whether or not human fetuses[12] are persons. It is a basic moral postulate that all persons have a full and equal right to life. The flip side of this postulate is that *only* persons have such a full and equal right to life. Thus, ending fetal life through abortion will be more difficult to justify if fetuses are persons than if they are not.

Yet the case for women's right to choose abortion does not rest entirely upon the claim that fetuses are not persons. If fetuses are persons, then deliberately induced abortion—insofar as it entails the fetus's death—must be seen as either a form of homicide or a case of deliberately allowing a person to die. As such, however, it may still be morally justifiable in many instances. It may, for instance, be a legitimate exercise of the right to self-defense.

The standard case in which killing in self-defense may be justified is that in which there is no other way to defend oneself against death or serious bodily injury at the hands of an unjust attacker. Abortion is different from this standard case of self-defense, in that the fetus is not *deliberately* endangering anyone, and in that childbirth today usually entails only a small risk of death. But there may be some cases in which individuals have the right to use lethal force in the defense of their own physical integrity, even though the person posing the threat is morally innocent, and even though the threat posed is not likely to prove lethal. (Consider, for instance, the case of self-defense against a rape attempt by an individual who is known to be mentally incompetent, and who never attempts to kill his victims.[13])

Abortion may also be justified as a refusal to act as an exceptionally good Samaritan. It may be argued that no one is morally obligated to allow her body to be severely altered and perhaps permanently damaged, in order to preserve the life of another person to whom she has not contracted any such special obligation.[14] No one may legally be forced, for instance, to donate a kidney or other body part, even to save another person's life. It may be selfish to refuse to make such a donation, but it would be seriously wrong to force an unwilling person to do so. Forcing a woman to complete an unwanted pregnancy is an outrage comparable to that of involuntary organ donation.

The standard reply to this argument is that women lose the right to self-defense against unwanted pregnancy when they voluntarily engage in heterosexual intercourse, with or without contraception. Many unwanted pregnancies occur in spite of the most conscientious use of contraceptives. The argument is that a fertile woman who, knowing this, fails to refrain loses the right to abort any resulting pregnancy.

But it is absurd to claim that engaging in heterosexual intercourse negates a woman's right to defend her physical integrity. Men who engage in heterosexual intercourse do not lose this right, nor should they. A father cannot legally be forced to donate a kidney, or even some easily replaceable blood, in order to save the life of his child. If men are not stripped of that right because of their sexual activity then it is doubly unjust that women should be. For, as Adrienne Rich has so eloquently argued, heterosexual intercourse is compulsory for women in ways it is not for men.[15] A majority of women are still forced in one way or another to market their sexuality to men in order to make a living—whether as wives, university professors, or nightly newscasters. Lesbians, like male homosexuals, are subject to severe social, legal and economic sanctions. A life of permanent celibacy is not socially or emotionally feasible for most women.

In short, compulsory heterosexuality makes nonsense of the claim that by engaging in sexual intercourse with men women voluntarily waive the right to defend themselves against unwanted pregnancy. Moreover, even if there were no such institution as compulsory heterosexuality, it would still be unreasonable to demand that heterosexual women remain celibate except at those times (if any) when they want to have a child. For most people, the expression of sexuality is far too important to reserve for just a few occasions in a lifetime. Granted, there are other ways of expressing sexuality than through the one act euphemistically referred to as heterosexual intercourse. But a majority of people seem to regard heterosexual copulation as at least one very important form of sexual expression, and it will not do to assume that it is something which anyone can easily give up, without serious loss.

If these arguments are sound, then women have the moral right to choose abortion regardless of whether or not fetuses are persons. Yet the issue of fetal personhood cannot be avoided. For if fetuses are persons, then abortion is at best a moral evil which is necessary under current conditions. If we believe that fetuses are persons then we must conclude that each society is morally obligated to make whatever changes in its institutions and practices are necessary to prevent the occurrence of unwanted pregnancies and consequent abortions. These changes will have to be very radical indeed. Until the day that a safe and 100 percent reliable means of contraception is invented and made universally available, it will be necessary to ensure that heterosexual intercourse never occurs between fertile persons unless the woman is willing to bear a child.

Such a severe restriction on sexual behavior may have often been mandated, but it has rarely been effectively enforced. Now, however, the effective enforcement of such a restriction is at least theoretically possible. For instance, all heterosexual men could be required to have vasectomies, with the option of having some of their sperm frozen for future use. But, so far as I know, no one has ever seriously advocated that any such measure be taken in order to reduce the need for abortion. Why *aren't* we willing to consider such measures, if fetuses really are persons with a right to life? If men could save the lives of millions of persons by submitting to a safe, simple surgical operation, wouldn't they be wrong (even if within their rights) to refuse to do so? Yet, rather than advocating such effective ways of reducing the need for abortion, the opponents of abortion propose only that it be legally prohibited—an expedient which history has shown to be as disastrous as it is ineffective.

This observation suggests that even the most radical opponents of abortion do not consistently act as though they believed that fetuses are persons with a full and equal right to life. It does not, however, help us decide whether *we* ought to believe that they are.

To state the case against this belief, it is necessary to make some general points about moral rights. Moral rights are protec- designed to safeguard the interests of the rights-bearer. This protec- tion consists in the moral prohibition of certain behaviors on the part of (other) moral agents. To say that persons have a right to life, for instance, is to say that it is morally wrong to deliberately or negligently take the life of any person—except in certain special cases such as justified self-defense. The right to life is a basic or especially important right because persons have an especially powerful interest in life, which is a precondition for the fulfillment of most of their hopes, plans, and desires. Where there is no interest in need of protection, talk of rights is not called for. For example, it would be quite pointless, and indeed false, to say that all people have a right to have a blue dot tatooed on their foreheads at birth, unless there were some reason to believe that having this done would somehow be in their interest. People can have rights only to things which are or tend to be in their interest. And, by the same token, only entities which can have interests can have moral rights.[16]

Interests are not the same as desires, since people may desire things which are not really in their interest (e.g., the alcoholic who wants another drink), or fail to desire things which are in their interest (e.g., the smoker who does not want to quit). Nevertheless, only beings who have some desires can be said to have interests, since only such beings can prefer—or have reasons for preferring—some states of affairs to others. Moreover, only entities which are capable of sentience (i.e., which can have experiences or feelings) can have desires. This is a perfectly common sense observation. We do not ascribe desires to tables, trees, or automobiles, except metaphorically, because we know that these things are not centers of experience:

they have no sensory experiences, cannot feel pleasure or pain, cannot be happy or frustrated, and therefore cannot literally be said to want anything. If only beings which have some capacity for sentience can have desires, and hence interests, then it follows that only such beings can have moral rights.

This argument does not, as it may first appear, have the absurd consequence that persons who are asleep or temporarily unconscious have no moral rights. To have moral rights, one need only have the *capacity* for sentience. The capacity for sentience requires only a brain and nervous system which are well developed and intact enough to permit the occurrence of conscious experiences. Persons who are asleep or temporarily unconscious have not lost that capacity. They may be incapable of having experiences at the present time, but they retain the capacity to have experiences later, without undergoing any further process of biological development or maturation. They still have desires (in the dispositional sense), even though they may not be currently aware of those desires. In contrast, a person whose brain has been permanently damaged such that no return to consciousness is possible has lost the capacity for sentience, and is, in many jurisdictions, legally defined as dead.

Such "brain-dead" individuals no longer have a moral right to life, for the simple reason that they have nothing to gain from continued biological existence, and nothing to lose from its cessation. For the same reason, we do not speak of rocks, trees, mountains, or other nonsentient elements of the natural world as having moral rights. They do not need moral rights, since they have no interests: there is nothing which can happen to them which can harm them in a way which matters to them. Trees and other nonsentient living organisms are goal-directed systems, and thus we can speak of some conditions as being good for them, that is, conducive to their health and survival. They can also "suffer," in the sense that their health may be impaired or their life span shortened. But such "suffering" does not entail, for them, any unpleasant experiences, or the frustration of any desires. (Desires can, of course, be frustrated without the occurrence of any unpleasant experiences; if this were not the case we would have difficulty explaining why even painless and unexpected death can be an evil.)

Of course, it is often necessary and appropriate to protect trees and other living organisms which are incapable of sentience. That an entity has moral rights is only one of many possible reasons for protecting it; other possible reasons are that it is valuable to us, or that its destruction might have damaging effects upon some ecological system. We may even find it useful to give *legal* rights to such organisms, as a practical means of assuring their protection.[17] The natural world must be protected from excessive human intervention, not just because our own survival depends upon the health of the global biological community, but also because it has what some philosophers have called "inherent value" or "inherent worth."[18] The homocentric perspective which holds that the destruction of a forest or the loss of a plant or animal species is bad if and only if human beings suffer as a result cannot be morally justified.

It is, nevertheless, important to maintain the distinction between those elements of the natural world which should be protected because of their inherent value, and those beings which should be protected through the ascription of moral rights. Moral rights function to override narrowly utilitarian considerations of value. All sentient beings may reasonably be held to have some moral rights, even though it is generally agreed that only persons—human or otherwise—have full and equal moral rights. For instance, any being capable of experiencing pain may be said to have a moral right that pain not be needlessly inflicted upon it.[19] It is wrong to torment even a fly for no reason other than one's own amusement. It is wrong not just because cruelty to animals can lead to cruelty to human beings (as Kant argued), but because it hurts the fly.

In the early stages of their development, human fetuses have no capacity for sentience, but only the potential to develop that capacity later. It is thought that a rudimentary capacity for sentience begins to emerge at some point in the second trimester of fetal development. By the third trimester some capacity for sentience is almost certainly present in the normally developing fetus.[20] It is unlikely, however, that any first-trimester fetus has even a rudimentary capacity for sentience, and it is quite certain that no fetus in the first half of the first trimester does. The evidence that first-trimester fetuses lack such a capacity is similar to and just as strong as the evidence that plants and some of the more simple forms of animal life are not sentient. They do not have brains or sensory organs which are sufficiently developed to permit the occurrence of conscious experiences, and neither does their behavior provide evidence of such experience. If first-trimester fetuses are not even minimally sentient, then they cannot have desires, or interests which could be protected through the ascription of moral rights. They are, to be sure, biologically human: they belong to the human species rather than some other species. But they are not yet beings, i.e., centers of experience.

These facts compel us to conclude that presentient fetuses do not yet have a right to life. In losing their possible future lives, they lose nothing which is of value to them. Thus, abortion cannot be viewed as a wrong against them. Morally speaking, fertilized human eggs, embryos and small fetuses are closely analogous to unfertilized eggs or spermatozoa, which may be valued for their potential role in the reproductive process but which (as everyone agrees) do not yet have moral rights.

This argument can be rejected only if it can be shown that there is some morally relevant difference between a fertilized human ovum and a viable unfertilized ovum, together with some viable human spermatozoa. But there is no such morally relevant difference. Both are alive, both are biologically human, and both have the potential, under favorable circumstances, to develop into a human being. The only differences which may appear to have moral relevance are (1) that the conceptus contains a complete set of chromosomes *within a single cell*; and (2) that the *probability* of its developing into a human being is apt to be much greater. Let us consider these differences more closely.

A fertilized human ovum, or zygote, contains a complete set of 46 chromosomes, in which is biologically encoded all of the genetic information necessary for the development of a new human being. In contrast, an unfertilized egg or a spermatozoon contains just half of this genetic complement. But an ovum *together with at least one spermatozoon* is also genetically complete. Why should it be thought to make a crucial moral difference that in the one case the genetic material is contained within a single living cell, whereas in the other case it is contained in two separate living cells?[21] We would not insist that the blueprint for a house did not yet exist, or that it has no value, simply because it had been drawn on two separate pages, needing only to be joined together. If all we need, to have a human person with full moral rights, is a complete genetic blueprint for such a person, then we must conclude that a person begins to exist at some point earlier than conception. The absurdity of that conclusion is a result of the absurdity of the premise: a genetic blueprint for a person is not yet a person, even though it may later develop into a person.

It is sometimes held that the human ovum should be regarded as becoming a person at the time of fertilization, because fertilization represents an enormous increase in the *probability* of its further development. Any normal and viable egg/sperm pair has the potential for developing into a new human being, but the probability of its actually doing so is, in most cases, very small. Once fertilization occurs—provided that it occurs within the body of a healthy woman—that probability becomes much greater. But this change in probabilities is morally irrelevant; for what changes is not the probability that a new person has already begun to exist, but rather the probability that a new person will come to exist later. Moreover, if the woman decides to terminate the pregnancy, then that probability may be just about as minute as it was prior to fertilization.

Thus, neither the appeal to potential nor the appeal to probability is sufficient to show that fertilized human ova or pre-sentient fetuses have any greater right to life than do ununited egg/sperm pairs. A first-trimester fetus is not and never has been a center of experience. Thus, it cannot sensibly be said to have moral rights. Its life may have great value for others, in that the birth of a child may be greatly desired; but it has not yet begun to value its own life. An entity's right to life cannot be based on the value which its life has for others—a value which may be either positive or negative. It is the value which people normally place on their own lives which underlies the moral right to life. The value which people place on one another's lives provides a powerful additional reason for respecting the lives of persons, but it cannot be the basic foundation of their right to life. If a person's right to life were based upon the value which other persons place upon her life, then some persons would have more right to life than others, and some might have no right to life at all.

It does not follow from the fact that presentient fetuses have no moral right to life that all actions affecting them are outside the scope of moral or legal concern. There are sound moral reasons for protecting fetuses from those kinds of harm which may infringe upon the rights of the persons they later become, and/or those of other involved persons. It is entirely appropriate, for instance, that a child born handicapped because of some drug which was prescribed for the mother should be able to sue for damages if the drug was prescribed negligently or without appropriate cautions. Negligent actions which cause children to be born mentally or physically handicapped are immoral, not because they violate the rights of presentient fetuses, but because they violate the rights of the children who will suffer because of those actions.

Because presentient fetuses do not yet have moral rights, and because women have the right to defend their own physical integrity, voluntary early abortion is not inherently wrong. It is not even prima facie wrong. That is, it does not require any special justification, such as financial hardship or the risk of damage to the woman's mental or physical health should she be forced to complete the pregnancy.

To claim that the voluntary choice of early abortion raises no serious moral issue is not to ignore the emotional trauma which such a choice may involve. The decision to abort or complete a pregnancy often constitutes a major turning point in a woman's life. She must choose one of two quite different futures for herself and her family. Other involved persons may have a right to have some input into her decision. But the final decision is hers, because it is her bodily integrity which is at stake. It is a decision which may occasion a great deal of grief and regret. Such emotions are important, but they do not entail that voluntary early abortion is morally problematic in itself.

If early abortion requires no special moral justification, then early abortion for the purpose of sex selection cannot be held to be morally objectionable because it involves the termination of a pregnancy. It may, of course, be objectionable for other reasons. If it is done for sexist reasons, or if it will cause predictable harms which outweigh the predictable benefits, then it may be wrong for these reasons. But it is not wrong simply because it involves abortion.

Early Abortion, Late Abortion and Infanticide

So far I have spoken only of abortions performed in the first trimester of pregnancy, before the fetus has developed a capacity for sentience. I have argued that such abortions are not in themselves morally problematic. Later abortion presents a somewhat different moral picture, for several reasons. First, the danger to the woman is somewhat greater, as is the physical and emotional trauma which she must endure. This consideration is probably irrelevant to a woman's right to choose late abortion, but it is important in determining the strictly medical indications for abortion. It is also highly relevant to the issue of when women can justifiably be asked or advised to (voluntarily) terminate a pregnancy for the sake of some social good, such as limiting population growth.

A second reason for regarding late abortion as morally problematic is that at some point in the second trimester the fetus probably begins to develop some capacity for sentience. At that point it is no longer simply a human organism, but a nascent human being.

Even a sentient fetus is not yet a person. It does not yet have a capacity for reason or self-awareness, or any of the other mental capacities which are distinctive of persons. It is these distinctive mental capacities which justify (if anything does) the presumption that more mature human beings are persons, whereas those nonhuman animals which lack such mental capacities are not; for it is only with respect to such higher order mental capacities that human beings can plausibly be held to be fundamentally different from most other terrestial animals.

Some would argue that the capacity to enter into reciprocal social relationships, e.g., to love and care for other persons, is more centrally definitive of personhood than are any such strictly mental capacities. Reciprocal social relationships are certainly vital to the lives and development of human persons. It is doubtful that a human being who was entirely deprived of all such relationships could ever become a person. But even on a definition of personhood which is based upon the capacity to enter into reciprocal social relationships, sentient fetuses are probably not yet persons, since their capacity for social interaction is at best extremely rudimentary. They may react to various stimuli (such as music, human voices, loud noises, or the mother's heartbeat), but it would be stretching a point to argue that this is evidence of a capacity to engage in the kinds of reciprocal social relationships which exist among persons.

Besides, it we were to accept a definition of personhood based on the capacity to enter into reciprocal social relationships then we would have to conclude that many animals which are not normally regarded as persons, such as cats and dogs, *are* in fact persons. Such animals can and often do enter into reciprocal social relationships with human beings, as well as with one another. If we wish to hold that such nonhuman animals are not necessarily persons, and do not necessarily have the same moral rights as human persons, then we must conclude that the capacity for reciprocal social relationships is not, in itself, a sufficient condition for being a person.

In other words, social relationships are essential to our existence as persons, but not all beings which have some capacity for social interaction are persons. Persons are social beings of a special type, who have a relatively extensive capacity to conceptualize, reason, and communicate with one another about their world, themselves, and their individual and collective futures. It is these capacities which enable and ultimately compel persons to recognize one another as beings with full and equal moral rights.

These reflections imply that even sentient fetuses are not yet persons. For the evidence that these capacities do not develop in the human individual prior to birth is quite conclusive. Some of this evidence is summarized in Michael Tooley's excellent book, *Abortion and Infanticide*.[22] Tooley provides an extensive examination of the behavioral and neurophysiological evidence concerning the mental capacities of newborn infants. The mental capacities of fetuses are certainly not more sophisticated than those of newborn infants. Thus, if neonates lack certain mental capacities then it is safe to conclude that fetuses must also lack these capacities. The behavioral evidence shows that neonates have very little capacity for learning, recognition, thought, conceptualization, or memory—far less, indeed, than most mature nonhuman vertebrates. Such capacities begin to develop fairly soon after birth, but they are not capacities with which we are born. In comparison to the young of many other mammalian species, human infants are born at an extremely primitive stage of physical and mental development.

The neurophysiological evidence is equally clear. The development of neuronal processes and synaptic connections, and the myelination of axons in the outer layers of the cerebral cortex (the probable primary locus of consciousness) is incomplete at birth to a degree which almost certainly precludes a capacity for thought. The mental capacities which neonates and late-term fetuses lack are precisely those which are generally taken to justify ascribing stronger moral rights to human persons than to (most) other animals.

But, although sentient fetuses are not persons, the act of deliberately killing them may still be prima facie wrong, or in need of moral justification in terms of particular extenuating circumstances. It may be argued that all sentient beings have at least some right to life, simply because all sentient beings have (or tend to have) an interest in their own continued existence. If this argument is correct, then it would seem that some rather compelling moral justification is required for the killing of any sentient being, even if it is not a person, and even if that killing is done painlessly.

Such an argument would require us to question the moral acceptability of such common practices as killing animals for food, when we could probably be just as healthy and happy as vegetarians, and using poison or lethal traps to rid our houses of rodents, when we could use live-catch traps and release the animals where they would do no harm. Perhaps it would still be possible to justify killing nonhuman animals in such cases, where (unlike in the abortion case) the basic moral rights of persons are not likely to be at stake, but it will be much more difficult to justify such killing than is commonly assumed. Nevertheless, the view that all sentient beings have some right to life is coherent and defensible—provided that it is true that all sentient beings tend to have an interest in continued life.

It is sometimes argued that sentient beings which are not persons cannot have any significant right to life, because they lack self-awareness and the ability to conceptualize their own future. Consequently, it is argued, they cannot value their own continued survival in the way that persons do.[23] This is a plausible reason for concluding that it would be irrational to ascribe to them the same full and equal right to life that persons have. Yet it is implausible to hold that such beings do not value their own lives at all. They frequently act as though they do. It is tendentious to insist that the death-avoiding behavior of (for instance) a mouse is merely a matter of "instinct"—a word which is often used to conceal the fact that we really do not know why a creature behaves as it does. If it is reasonable to believe that some or all sentient beings which are not persons nevertheless value their lives, then there is room to argue that they have some right to life. On this theory, killing a sentient being which is not a person is not murder, but even mice should not be killed for no good reason.

A third reason for regarding late-term abortion as morally problematic is that a late-term fetus is not only a sentient being, but a sentient *human* being. Membership in the human species is not, by itself, a morally significant property. It does not provide an adequate basis for ascribing moral rights to presentient fetuses, or to mature human beings who have suffered irreversible brain damage which makes any return to consciousness impossible. Genetic humanity is not a sufficient condition for having moral rights. It is not a necessary condition either. Sentient, reasoning, self-aware beings which belonged to some other animal species (or sentient, reasoning, self-aware robots or androids) would be persons, and would have the same basic moral rights as human persons. If dolphins, whales, or apes are persons—and it is possible that they are—then the failure to recognize their moral rights as equivalent to our own is a prejudice which is comparable to sexism or racism. Peter Singer has dubbed this form of prejudice "speciesism."[24]

There may, nevertheless, be sound moral reasons for ascribing stronger moral rights to sentient human beings who are not yet persons than to comparably sentient nonhuman beings. Few people doubt that the killing of a newborn infant is harder to justify than is the killing of a mouse. Yet the mouse is probably more highly sentient—e.g., capable of more sophisticated perceptions and mental processes. Is it only speciesist discrimination which lies behind the conviction that killing babies is a more serious matter than killing mice? I do not believe that it is.

In classifying infanticide as a form of homicide, the law goes far beyond what can be justified on the basis of the intrinsic properties of neonates, who are less highly sentient and less self-aware than many nonhuman beings which receive far less protection. Many of the reasons behind the traditional Christian condemnation of infanticide are subject to no empirical verification. For instance, the infant may be thought to have an immortal soul, whose future will be jeopardized if the infant dies before earning a place in heaven through faith and good works. Such reasons can only be accepted on faith, and have no force for those who do not share that faith. There are, however, reasons for regarding infanticide as morally undesirable which do not depend upon the acceptance of any particular religious creed.

Pragmatic considerations can never justify denying full and equal moral and legal rights to any person. But it does not follow from this that it is wrong to extend additional moral and legal protections to some beings who are not persons, on the basis of sound moral and practical reasons. It is a plausible moral principle that sentient postnatal human beings whose mental development is so incomplete or impaired that their personhood may be doubted should nevertheless be treated as having the same basic moral rights as the rest of us, so long as this can be done without violating the basic moral rights of persons.

One good reason for accepting this moral principle is that the empathy and affection which most of us feel towards infants or severely retarded persons make us want to extend strong protections to them. This desire is not irrational. It is based on care and compassion, and these are valuable qualities. In the absence of very strong reasons to the contrary, this motivation to protect helpless human beings ought to be respected. Another good reason is that we know that we or those we care about may someday lose the mental capacities which are distinctive of persons—without at the same time permanently losing the capacity for sentience—and we want to ensure that our or their interests will be protected should this occur.

We must also recognize the often self-serving tendency of human beings to ignore or deny the mental capacities of those who are different—in sex, race, species, age, or some other variable—from what is taken to be the norm. Because we recognize this human tendency, we ought to give other sentient beings the benefit of the doubt in cases where there is legitimate uncertainty about whether or not they are persons. There is no uncertainty at all about whether fetuses are persons, since their development is far too incomplete to make this even a remote possibility. But there is some uncertainty about the mental capacities of very young infants, even though the currently available evidence strongly indicates that they are not yet capable of thought or self-awareness.

In the absence of decent foster care for infants whose parents cannot care for them, the prohibition of infanticide causes more harm than good. It contributes to poverty and the enslavement of women, and forces parents to resort to covert ways of eliminating "excess" children. These covert forms of infanticide are often much more cruel than the ancient practice of exposing unwanted neonates. On the other hand, where adequate foster care is available, the prohibition of infanticide need not have such disastrous results. It is only in the present century that more effective means of contraception and safer means of abortion have made it pragmatically possible to reduce the number of unwanted infants enough that the virtual elimination of infanticide can become a feasible goal. It is a goal which we ought to pursue, so long as we do not in the process violate the rights of parents and infants themselves.[25]

The key question, then, about the moral status of late abortion is whether the strong protections extended to newborn infants should also be extended to second- and third-trimester fetuses. In most contemporary legal systems, infants are classified as persons, and infanticide as a form of homicide. But even in jurisdictions in which infanticide has been classified as a form of homicide, late abortion has very seldom been classified in the same way. Most legal systems which are influenced by English common law treat the moment of live birth as the beginning of a person's legal existence.

Birth is not merely an arbitrary line of demarcation drawn by lawmakers. It is a morally significant event, because it marks the point at which the infant begins to exist as a biologically separate organism, no longer contained within the body of another. The point is not that the fetus is dependent upon the woman for its continued survival; that may continue to be the case long after it is born. The morally crucial difference between a sentient fetus and a newborn infant is not its degree of dependence, but its location. Normally, a being's location makes no difference to its moral status; but this case is unique. So long as the fetus remains within the woman's body, it is impossible to treat it as if it were already a person with full and equal moral rights, without at the same time treating the woman as if she were something less.

Why not? Why can't we treat both the woman and the sentient fetus as having full and equal basic moral rights, and simply weigh the two sets of rights against each other when they happen to conflict, just as we do in many other instances in which the rights of different individuals come into conflict? Abstractly stated, such a moral policy may sound both just and equitable. In practice, however, to classify sentient fetuses as persons would be to endanger virtually all of women's most basic moral rights.

Consider some of the probable results of such classification. All late abortions, whether spontaneous or induced, would be liable to be investigated as possible criminal homicides. Women and physicians could be charged with murder or manslaughter as a result of any late abortion performed for what someone else regarded as inadequate reasons. In the case of spontaneous miscarriage, virtually every aspect of a woman's private life might be subjected to scrutiny in order to determine whether the miscarriage might have been due to some negligence on her part. Her diet, sleep patterns, sexual behavior, exercise regimen, drinking habits, stress levels at work or play, indeed just about anything she did or did not do while pregnant (or even before becoming pregnant) might be used as a basis for a charge of manslaughter. Extreme as these consequences are, they could not logically be avoided if fetuses were legally classified as persons in the fullest sense.

Another, and particularly frightening, risk inherent in the suggestion that sentient fetuses should be treated as persons in the full legal and moral sense is that pregnant women may be forced, against their will, to undergo surgical or other medical interventions which are judged by others to be necessary for the health or survival of the fetus. There have already been several recent cases in which American courts have ordered pregnant women to submit to Caesarean sections, in order to protect the fetus from the supposed risks of vaginal delivery.[26] Other cases have established that parents do not have the legal right to refuse life-saving medical treatment for their children.[27] Thus, the danger exists—even if fetuses are not given any additional legal status—that pregnant women may be coerced into undergoing more and more invasive, dangerous and painful medical interventions, as the medical treatment of fetuses in utero becomes feasible in a wider range of cases.

For such reasons as these, it is impossible to extend the same moral and legal status to fetuses as to infants, without endangering women's rights to an intolerable degree. There is room for only one being with full and equal rights inside a single human skin. So long as infants are born of women, rather than gestated in artificial wombs, it will be impossible to treat both women and fetuses as having full and equal rights; and since women are persons and fetuses are not, it is women's rights which ought to prevail.

These considerations help to explain the common conviction that although late abortion is a morally more serious act than either contraception or early abortion, it is not as serious as infanticide. A strong case can be made that women have the right to seek abortion at any stage of pregnancy, regardless of whether or not fetuses are regarded as persons. But when we add to this the fact that fetuses are clearly not yet persons, the case for the right to choose abortion becomes still stronger.

Is Sex Selection an Acceptable Reason for Abortion?

If, as I have argued, early abortion raises no serious moral issues, then early sex-selective abortion is morally no more problematic than preconceptive sex selection. Furthermore, because even late-term fetuses are not yet persons, and because all persons have a basic moral right to "control their own bodies,"—i.e., to defend, and make decisions affecting, their own physical integrity—a woman has a moral right to choose even late abortion for any reason which she regards as sufficient. Thus, she has a moral right to use even late abortion as a means of sex selection. Of course, she also has the right not to. No one else has the moral right to coerce her in either direction.

It is important to realize, however, that one may sometimes act within one's rights, and yet not be doing the right or morally optimal thing. Unlike early abortion, late abortion requires some morally sound justification. In other words, it requires rather sound reasons to show that it is not merely something which the woman has the right to choose, but which she is right to choose. Protecting her own life or health or avoiding the birth of a severely handicapped infant are widely, and I think rightly, accepted as adequate justifications for late abortion. That sex-selection is an adequate reason for late abortion is, however, much less clear.

I believe that it is morally objectionable to deprive a sentient human being of its life because of its sex. A late-second- or third-trimester fetus is not yet a person, but it is probably already a center of experience. Thus, it is not merely a human organism, but an incipient human being. The abortion of second-trimester male fetuses which are likely to suffer from some severe sex-linked disease is different from sex-selective abortion in that the fetus is not killed because of its sex, but in order to prevent the coming into being of a person who, because of severe and incurable illness, is likely to have little or no chance to lead a happy life. There may be other objections to abortions performed for this reason,[28] but such abortions are not an expression of sexism.

In contrast, the abortion of second- or third-trimester fetuses because they are female is an extreme expression of the wrongful devaluation of all females. It need not reflect sexist attitudes on the part of the individual parents, but it is certainly a reflection of sexism on the part of the society. It requires a stronger reason than sex-preference to justify the killing of a sentient human being—even one which is not yet a person. Sexist social institutions can create powerful pragmatic reasons for son-preference, reasons which may seem compelling even to individuals who hold no sexist beliefs. But there is no justification for the patriarchal institutions which lead to the devaluation of females and the deliberate killing of sentient female human beings. In highly sexist societies, sex-selective abortion or infanticide may sometimes be the lesser evil; but it is an evil nevertheless. In those rare instances in which sentient fetuses may be aborted simply because they are male, the moral objection still applies. Thus, sex-selective late abortion is an instance of gendercide, even though it is not an instance of murder.

Yet before we conclude that moral censure is an appropriate response to those who use late abortion for sex selection we should consider the extreme severity of the action for the woman herself, and the desperation which must be necessary to impel a woman to make or consent to such a choice. Caroline Whitbeck points out that: "To have a third trimester abortion for trivial reasons would be an act of major self-mutilation, as anyone who understands the experience of pregnancy will realize."[29]

Very few women would submit to a late-term abortion for trivial reasons. However, in a highly patriarchal society, sex selection may *not* be a trivial reason for abortion. It may be an extremely compelling reason, from the viewpoint of the individuals who must make the decision. Under some circumstances, having a child of the "wrong" sex may be a misfortune of major proportions, and a female child, if born, may have little prospect of leading a decent life. In such circumstances, the blame must attach not to the individuals who opt for sex-selection abortion, but to the society which has created these circumstances. We should feel only compassion for women who find it necessary to accept such an extreme measure in order to prevent the birth of an unwanted daughter. If we want to prevent sex-selective late abortion, we must change the patriarchal attitudes and institutions which can make such a choice (seem to be) the lesser evil.

Notes

1. See Janet Radcliffe Richards, *The Sceptical Feminist: A Philosophical Inquiry* (Middlesex, England: Penguin Books, 1983), 74–75. Richards provides a lucid

discussion of the problems encountered in interpreting the distinction between what is natural and what is unnatural in human behavior, and in using any such interpretation as a basis for drawing moral conclusions.

2.   Simone de Beauvoir, *The Second Sex*, translated and edited by H.M. Parshley (New York: Random House, 1974), 58.

3.   President's Commission for the Study of Ethical Problems in Medicine and Biomedical and Behavioral Research, *Screening and Counseling for Genetic Conditions* (Washington, D.C.: U.S. Government Printing Office, 1983), 58.

4.   Ibid.

5.   Tabitha Powledge, "Unnatural Selection: On Choosing Children's Sex," in *The Custom-Made Child? Woman-Centered Perspectives*, edited by Helen B. Holmes, Betty B. Hoskins and Michael Gross (Clifton, New Jersey: Humana Press, 1981), 197.

6.   Michael Bayles, *Reproductive Ethics* (Englewood Cliffs, New Jersey: Prentice-Hall, 1984), 35.

7.   John Fletcher, "Is Sex Selection Ethical?" *Research Ethics* (New York: Alan R. Liss, 1983), 347.

8.   See Bernard Horowitz and Madhu Kishwar, "Family Life—the Unequal Deal," *Manushi* 2 (1982), 2–18.

9.   Fletcher, 343.

10.   Letty Cottin Pogrebin, *Growing Up Free: Raising Your Child in the 80's* (New York: Bantam Books, 1981), 100.

11.   Sally Gearhart, "The Future—if There Is One—Is Female," *Reweaving the Web of Life: Feminism and Nonviolence*, edited by Pam McAllister (Philadelphia, Pennsylvania: New Society Publishers, 1982), 282.

12.   Technically, a conceptus is referred to as a fetus only after all of the major organ systems are formed, at about eight weeks of gestational age. For simplicity, however, the term "fetus" is often used to refer to a conceptus at any stage of development.

13.   Jane English imagines a case in which innocent persons are "programmed"by an evil scientist to assault other innocent persons with knives. She points out that if one were to be attacked by one of these innocent victims, and if killing the latter were the only way to avoid being seriously injured, then this would be justifiable—even if one knew that it was unlikely that the damage one would otherwise suffer would prove to be lethal. ("Abortion and the Concept of a Person," *Social Ethics*, edited by Thomas Mappes and Jane Zembaty (New York: McGraw-Hill, 1982), 30–37.)

14.   Judith Jarvis Thomson develops this line of argument in her classic article, "A Defense of Abortion," *Philosophy and Public Affairs* 1(Fall, 1971), 47–66. Thomson imagines a case in which you are kidnapped and your bloodstream hooked up to that of a famous violinist. The violinist has an illness which will certainly kill him unless he is permitted to make use of your kidneys for a period of nine months. In such a case, she argues, you would surely have the right to unhook yourself from the violinist. Even though the latter would die as a result, you would not be guilty of murder, or any other moral offense. In no other case than abortion, she notes, has the law required individuals to act as good Samaritans at such enormous cost to themselves.

15.   Adrienne Rich, "Compulsory Heterosexuality and Lesbian Existence," *Signs: Journal of Women in Culture and Society* 5:4 (Summer, 1980), 631–60.

16.   Joel Feinberg defends this claim, which he calls the interest principle, in "The Rights of Animals and Unborn Generations," in *Philosophy and Environmental Crisis*, edited by William T. Blackstone (Athens: University of Georgia Press, 1974), 51.

17.   See C.D. Stone, *Should Trees Have Standing? Toward Legal Rights for Natural Objects* (Los Altos, California: William Kaufman, 1974).

18.   See, for instance, Tom Regan, *All that Dwell Within: Animal Rights and Environmental Ethics* (University of California Press, Berkeley, 1982), 203; and Paul

Taylor, "The Ethics of Respect for Nature," *Environmental Ethics* 3 (1981), 197–218.

19. To speak of the rights of a fly is somewhat counterintuitive. The moral wrong committed in tormenting a fly may not seem significant enough to justify describing it as a violation of a right. The argument for speaking of rights is that even a fly has interests which require some moral consideration—even though the obligation to respect those interests is not nearly as powerful as is the obligation of respect for the interests of persons.

20. See L.W. Sumner, *Abortion and Moral Theory* (Princeton, New Jersey: Princeton University Press, 1981), 149.

21. R.M. Hare makes a similar point in "Abortion and the Golden Rule," *Philosophy and Public Affairs* 4:3(1975), 212.

22. Michael Tooley, *Abortion and Infanticide* (London: Oxford University Press, 1983), 347–407.

23. See Ruth Cigman, "Death, Misfortune, and Species Inequality," *Philosophy and Public Affairs* 10 (Winter, 1981), 48.

24. Peter Singer, *Animal Liberation* (New York: Avon Books, 1975), 7.

25. In relatively wealthy societies, the absolute prohibition of direct or indirect infanticide becomes problematic primarily in cases where the infant is extremely premature, or suffers from congenital deformities severe enough to cast doubt on the presumption that prolonging its life by every available means constitutes a defense of its interests. Infants who only a few years ago would have had no chance of survival can now be kept alive much longer. Many of these have no prospect except a very short life, or one burdened by severe disabilities. Where reasonable persons may disagree about what medical treatment best serves the child's interest, the decision to pursue or withhold particular treatments is best left to the persons who are most directly involved—the parents and their medical advisors.

26. George J. Arras, "Forced Caesarians: The Unkindest Cut of All," *Hastings Center Report* 12:3(June 1982), 16–17.

27. These include cases in which parents who are Jehovah's Witnesses, and who believe that the use of blood transfusions is contrary to biblical injunctions, have been required to permit such transfusions to save their children's lives.

28. See Chapter 7, Part 1.

29. Caroline Whitbeck, comment during audience discussion at conference entitled Human Life Symposium: An Interdisciplinary Approach to the Concept of Person, *Defining Human Life; Medical, Legal, and Ethical Implications*, edited by Margery W. Shaw and A. Edward Doudera (Ann Arbor, Michigan: AUPHA Press, 1983), 221.

5
More Males/More Violence?

I have argued that sex selection through preconceptive methods or early abortion is not intrinsically wrong. That is, it is not wrong simply because of the kind of act it is, and apart from its probable consequences. However, if the probable ill effects of the use of new methods of sex selection greatly outweigh the probable benefits, then it may be wrong for utilitarian reasons. The possible benefits—such as reducing the birth rate and the number of children born unwanted—will be discussed in Chapter 7. This and the following chapter will deal with the undesirable consequences which some have predicted from the use of new methods of preselecting sex. Perhaps the most alarming of these predictions is that by increasing the relative number of males, sex selection will lead to a more violent world.

The Aggressive Sex

One of the most consistent cross-cultural differences between women and men is that men are more apt to behave violently. The vast majority of violent crimes are committed by men. In the United States, males commit approximately five times as many homicides and twenty times as many armed robberies as do females.[1] Rape and the sexual abuse of children are predominantly male crimes—though there have recently been many cases in which women have been accused of sexually molesting children, and one in which two women allegedly conspired to rape a man who had jilted one of them. In all societies which engage in organized warfare it is primarily men who are warriors. Even the famous female warriors of the African kingdom of Dahomey were outnumbered by male warriors. Roy D'Andrade analyzed sex-role differences in over 600 cultures, and concluded that in virtually every instance, males are more aggressive and more dominant than females.[2] Maccoby and Jacklin's analysis of research on sex differences done between 1966 and 1973 shows that one of the most consistent findings is the greater aggressiveness of males at all ages. While there are many women who are more aggressive than many men, nearly every study has shown that the average male is significantly more aggressive than the average female from the same cultural background.[3] This difference is most apparent with respect to overt physical aggression against other persons. As a group, females occasionally come out ahead in measurements of verbal aggression, but never in measurements of physical violence.

The greater propensity of males to resort to violence has led some to argue that increased sex ratios will lead to a more violent society. Amitai Etzioni speculates that,

> A significant and cumulative male surplus will ... produce a society with some of the rougher features of a frontier town. And ... the diminution of the number of agents of moral education and the increase in the number of criminals would accentuate already existing tendencies which point in these directions, thus magnifying social problems which are already overburdening our society.[4]

Of course, these are stereotypes: men are violent, women are agents of moral education. But not all stereotypes are wholly false. If Etzioni's speculation is plausible, then it constitutes a powerful argument against the use of sex selection in son-preferring societies. The problem of human violence has become acute in our time. We are the generations who "walk through the valley of the shadow of nuclear annihilation, and we do fear evil."[5] We cannot afford a worldwide outbreak of military hostilities. Nor can we accept the high levels of domestic and street violence in our own and many other societies. "With our present state of technological development, uncurbed violence may mean the disintegration of social structure, nuclear holocaust, and collective suicide."[6]

But is it true that the use of the new methods of sex selection is likely to lead to an increase in violence? Before attempting to answer this question, we need to clarify what is meant by aggressiveness, that trait which is so much more marked in males. Is aggressiveness neccessarily a bad thing? Next, we need to consider the evidence that the greater aggressiveness of human males is at least in part a result of male biology, and not just of culture and social learning. If male aggression is an inevitable result of male biology, then more males will inevitably mean more violence. But if cultural and environmental factors are more important determinants of male aggression, then it is not obvious that higher sex ratios will mean more violence.

Defining Aggression

The term "aggression" is notoriously vague, and can be used either to praise or to condemn. We praise football players for their aggressive play, while condemning hostile nations for their "acts of aggression." Webster's defines aggression as "an unprovoked attack," thereby excluding even the most violent behavior, if it occurs in response to some provocation.[7] In contrast, aggression is sometimes defined as any behavior that leads to increased social status.[8] On this definition, a person who is quietly meditating or reading a book may be said to be behaving aggressively, if the society happens to reward such behavior with increased status.

For our purposes, the narrower definition provided by Webster's is more appropriate. Males tend to be more aggressive than females primarily in that they are more apt to make unprovoked attacks on others. Such behavior does not necessarily lead to increased social status. Nor are all behaviors that enhance social status by definition aggressive; if nurses had high social status, that would not make nursing a form of aggression.

The Webster's definition is, however, somewhat too narrow. Aggression includes not only the making of unprovoked attacks, but also behavior which is hostile and threatening. The qualification "without provocation" also needs to be clarified. Aggressive individuals may not become hostile except when they believe that they have been provoked. However, their "provocation threshold" will be lower than that of less aggressive individuals. We would not normally describe a person who strikes a mugger in self-defense as behaving aggressively, because the response is not disproportionate to the provocation. But a person who strikes another for making a mildly critical remark would be described as aggressive, because the provocation is minor in comparison to the response. Aggression, then, is threatening or assaultive behavior, in response to little or no genuine provocation.

Aggression is sometimes said to be a good thing. Some argue that it is: "a useful human trait, if appropriately handled. We all need to be properly aggressive and society needs aggressive individuals to carry out programs of social reform."[9] However, in the sense of the term which I am employing, aggression is almost never socially desirable. Sometimes it is necessary to fight in defense of oneself or others, and sometimes it is appropriate to pursue legitimate goals vigorously, and in defiance of all opposition. But these are not acts of aggression. Aggression may sometimes benefit the aggressive individual or society; but it is usually morally objectionable. It also creates resentment and invites eventual retaliation. The common belief that aggressiveness is associated with superior creativity or other socially desirable traits is unproven and quite implausible. There are as many mild-mannered geniuses as belligerent ones.

The Universality of Greater Male Aggressiveness

If the greater aggressiveness of males is a cross-cultural universal, then this universality is prima facie evidence that it is rooted in male biology. But *is* male aggressiveness a cross-cultural universal? In one sense it is, but in another it is not. In virtually every known culture, males tend to be somewhat more aggressive than females of that culture. Yet there is enormous variation among different cultures with respect to the amount of aggressiveness displayed by members of either sex.

We have already met the Arapesh and the Mundugumor, two Papuan cultures described by Margaret Mead. The contrast between these societies is often cited as evidence of the amazing malleability of human nature; and so it is. But there are also some interesting similarities between the two cultures which suggest that some aspects of human nature are not simply a function of culture, and that the greater aggressiveness of males may be one of these. The Arapesh are a nonmilitaristic people who abhor most forms of aggression. Their culture, Mead reports,

> substitutes responsiveness to the concerns of others, and attentiveness to the needs of others, for aggressiveness, initiative, competitiveness and possessiveness—the familiar motivations upon which our culture depends.[10]

Among the Arapesh, there is said to be very little sex stereotyping of personality or temperament. The expectation is that both sexes will be unaggressive, unacquisitive and cooperative. Violent individuals of either sex are looked upon with disfavor. Nevertheless, some violence occurs, and it is most often committed by men. Although there is no organized warfare, there are occasional skirmishes between men of different villages, often due to conflicts over women or livestock. The opposing groups hurl spears at one another, with the intention not of killing but of wounding slightly; the battle ends immediately if someone is badly hurt.[11] Women do not engage in such skirmishes, nor are they as apt to commit violent attacks upon members of their own community. It is also the men who hunt.

The Mundugumor, as Mead describes them, are very different. The men of this tribe were headhunters in the past, and that practice had only recently been suppressed by the British colonial government at the time Mead made her study. Among the Mundugumor, the normal temperament for members of both sexes is violent, jealous, competitive, and vengeful. Relationships between members of each sex are characteristically hostile, and even heterosexual affairs tend to be rather violent. Husbands often beat their wives; but the latter may fight back, and the men sometimes get the worst of the encounter. Men may have several wives. Since wives bring wealth through their productive activities, the competition for women is intense. Co-wives are jealous of one another and of their daughters. Children are reared without much solicitude, learning to fight for survival from the earliest age. But although there is no expectation that women will be more gentle than men, nevertheless it is the latter who most often engage in the more extreme forms of violence, such as raiding neighboring villages to kill or take captives. Within the community, hostilities between women are less apt to explode into deadly combat.[12] Evidently the sanctioning of aggressive behavior on the part of both sexes—like the condemnation of such behavior for either sex—is not sufficient to produce a society in which the average man is no more aggressive than the average woman.

Why do males tend to be more aggressive than females of the same cultural background, even in the absence of any apparent sexual stereotyping of aggressiveness? Even in the case of the Arapesh and Mundugumor, culture undoubtedly plays a role in encouraging greater male aggressiveness. Although neither society has dichoto-mized concepts of masculine and feminine temperament, distinct sex roles do exist. It is by custom, and not merely a matter of individual choice, that women do not hunt or take up arms against neighboring peoples. Among the Arapesh, the expression of anger is discouraged earlier in girls than in boys.[13] Arapesh men are expected to exercise authority over their wives. Girls are often married while they are still children, and are dependent upon their husbands for their subsistence until they reach maturity. It is men who exercise extrafamilial authority, through the role of the "big man" who plans public feasts and exchanges of food. This role is considered more a burden than an honor, and is undertaken with reluctance and surrendered with relief; but there are no "big women" with comparable authority.

Nevertheless, the influence of culture is insufficient to explain why males are the more violent sex in virtually every human culture. Most cultures—including our own—are more tolerant of aggressive behavior on the part of males than on the part of females; but this cross-cultural tendency may itself be rooted in biological differences between the sexes.

The Influence of Male Hormones

It has often been suggested that the male hormones, or androgens, are responsible for the propensity of human males towards violent aggression. These hormones are present in persons of both sexes. However, higher androgen levels occur in postpubertal males, and in the male fetus in certain stages of development. The physiological effects of testosterone, the most important of the androgens, are known to account for many observable physical differences between the sexes, including differences in genitalia and muscular development.

Much less is known about the psychological and behavioral effects of sex hormones in the human species. These effects have been most thoroughly studied in such laboratory species as rats, mice and rhesus monkeys. In these and other mammalian species which have been studied, testosterone has been found to play a dual role in the production of characteristically masculine behavior, such as rough-and-tumble play (in the prepubertal state), and (after puberty) mounting, and fighting—especially with other males. First, testosterone has been found to act upon the brain of the normal male during an early critical period, thereby predisposing the individual towards masculine behavior patterns. Second, it plays an activating role in later life, eliciting aggression and masculine mating behavior in individuals that have been exposed to androgens during the early critical period.

In mice and rats, the critical period is shortly after birth. Male mice castrated at birth fight less than intact males, even when given later testosterone injections. Female mice which are given testosterone at birth and additional injections at puberty develop masculine mating behaviors, and fight about as much as males. Without neonatal exposure to testosterone, later testosterone injections will not induce such behavior.[14] In rhesus monkeys, the critical period is prenatal. Injecting pregnant females with testosterone up to 85 days prior to the birth of the offspring results in females with masculinized genitalia. When immature, these masculinized females engage in more rough-and-tumble play and more threatening and chasing behavior. After puberty, they fight more than normal females and more often mount other females.[15]

These and other experiments have established that the greater aggressiveness characteristic of most male mammals is in large part due to the action of testosterone, and possibly other steroid hormones. These hormones appear to "have the ability to alter or modulate the level of excitability in adult brain structures that control the expression of aggression."[16] Yet the propensity for aggressive behavior is also affected by environmental conditions, even in nonhuman mammals.

Some popular writers who argue that biology inevitably predisposes human males towards aggression cite as evidence the behavior of the savanna-dwelling hamandryas baboons.[17] According to most reports, the males are ferocious bullies who dominate the much smaller females and maintain strict dominance hierarchies among themselves. But some observers have noted that the behavior of forest-dwelling baboons of the same species is very different. There, aggression is less frequent, and male dominance hierarchies are less marked.

David Pilbeam suggests that the behavior of the baboon troops first studied was in a sense abnormal. These troops were living in the game parks of Kenya and Tanzania, where food and cover are scarce, predators numerous, and human interference frequent. These factors probably produce an abnormally high level of stress, which may explain the highly aggressive behavior of the males.[18] Male baboons are more prone to aggression than females, but it seems that testosterone does not simply cause aggression, independently of the environmental conditions. Instead, "the effects of tetosterone must be regarded as acting permissively, allowing the elicitation of an attack by social stimuli, and set within a framework of modification of the CNS (central nervous system) by past experience."[19]

Analogies between the behavior of human and nonhuman mammals are insufficient to establish the operation of similar causal factors in both cases. Human behavior is influenced by conscious moral commitments, and other intellectual processes which are made possible by the use of human languages. It is impossible to prove conclusively that any differences between the aggressiveness of male and female human beings are due to biology rather than to culturally inculcated sex roles. For there is no human culture which is entirely free of sex roles. As John Stuart Mill pointed out, only given such a culture would it be possible to determine whether there are any innate differences between the psychology of men and women.[20]

There is, nevertheless, some evidence that the differences between the behavior of human males and females may be due in part to the psychological effects of the male hormones. Some of this evidence comes from studies of individuals suffering from hormonal abnormalities.

Perhaps the most important of these studies are those done by Anke Ehrhardt and John Money at Johns Hopkins University and the Pediatric Endocrine Clinic in Buffalo. Ehrhardt and Money studied the behavior of children born with the adrenogenital syndrome (AGS). AGS is a rare condition which occurs in both females and males. Beginning at an early stage of fetal development, the adrenal glands produce abnormally large amounts of testosterone. Girls born with AGS usually have fairly normal internal reproductive organs, but somewhat masculinized external genitalia. The standard treatment for girls with AGS involves surgical correction of the genitalia and lifelong hormone treatment, to prevent further physical masculinization. Boys with AGS undergo abnormally early puberty if they are not also given corrective hormone treatments.

In the early 1970s, Money and Ehrhardt conducted interviews with fifteen AGS girls, and with their parents, siblings, and teachers. They reported that all of these girls considered themselves to be female, and none expressed any interest in becoming male. However, they seemed to be more interested in rough outdoor play than are most girls in our culture, and less interested in dolls than in toy trucks, guns, and other "boys' toys." They also tended to prefer utilitarian to "feminine" clothing, and to express less interest in marriage and motherhood than in professional careers.[21] Ehrhardt's later study, which compared groups of AGS girls and boys with their unaffected same-sex siblings, produced similar results. Both the AGS girls and the AGS boys tended to have a higher level of energy expenditure and a greater interest in rough outdoor activities than did siblings of the same sex. The AGS girls were more apt to prefer boys as playmates and to describe themselves as tomboys. They were less concerned about their appearance and less interested in dolls than those in the control group.[22]

Erhardt has also studied a group of girls who were fetally androgenized as a result of exposure to progestinic drugs. During the 1950s these drugs were administered to pregnant women in the erroneous belief that this would help to prevent miscarriage. These girls, like the AGS girls, were born with somewhat masculinized external genitalia. They were treated only with surgical correction, since their postnatal hormone levels were not abnormal. Once again, the familiar pattern of "tomboyish" behavior was found.[23] Interestingly, however, neither the AGS boys nor the two groups of fetally androgenized girls were found to be significantly more aggressive than those in the control groups. Children in the former groups were said to be slightly more likely to start fights, but the difference fell below the level of statistical significance. The groups were quite small (between 10 and 20), and it is possible that had they been larger a significant difference in aggression would have been found. Nevertheless, this negative finding does suggest that, while fetal exposure to high levels of androgen may have some effect upon later behavior, it does not have a direct effect upon levels of aggression.

Money and Ehrhardt have described another case which also seems to support this conclusion. One of a pair of identical twin boys was accidentally deprived of his penis while being circumcised, at the age of seven months.[24] The parents agreed to have the rest of the child's male genitalia removed, and to raise the child as a girl. The mother placed great emphasis on teaching her the female sex role, e.g., by allowing her hair to grow, dressing her in feminine clothes, and providing her with dolls and other "girls' toys." She reported that the "girl" twin accepted herself as a female and had no doubts about her sexual identity. Nor was she particularly aggressive, although she was said to have "many tomboyish traits, such as an abundant physical energy, a high level activity, stubbornness, and being often the dominant one in a girls' group."[25].

Attempts to establish correlations betwen current androgen levels and violent behavior in biologically normal males have not been particularly successful. Negative findings have been at least as frequent as positive correlations. Significant correlations between current androgen levels and aggressive behavior are more often found in young postpubertal males than among older men. For instance, Persky et al. measured testosterone levels in a group of eighteen young men and gave them a battery of written tests designed to measure verbal and physical aggressiveness, irritability, and resentment. In this group, a significant correlation was found between the rate of testosterone production and test measurements of these traits, whereas no significant correlation was found in a second group consisting of older men.[26]

Another interesting finding comes from a study by Kreuz and Rose of male prisoners, aged 19-32, in the Patuxant Institution at Jessup, Maryland. Levels of plasma testosterone were measured daily over a two-week period, and compared with the individual's history of fighting in prison and with the results of psychological tests designed to measure aggressiveness, hostility, and anxiety. These comparisons yielded no significant correlations. Furthermore, these young men did not have testosterone levels significantly higher than those observed in other groups of young men outside of prison. However, the ten prisoners who had committed particularly violent crimes in adolescence were found to have higher testosterone levels than the eleven prisoners incarcerated for less violent crimes, such as burglary.[27] The authors note that a possible explanation of this finding is that,

> in individuals predisposed to antisocial behavior by virtue of familial and social factors, increasing levels of testosterone during adolescence serve to precipitate such behavior ... testosterone may serve to stimulate increased activity, drive or assertiveness, and in certain individuals this may be utilized in antisocial, aggressive acts.[28]

But such findings can never prove definitively that any of the differences between male and female behavior are due to the effects of testosterone. For it is impossible to eliminate other possible explanations.

There are, for instance, alternative explanations of the findings of Ehrhardt and Money. It may be that the so-called "tomboyish" behavior of the two groups of fetally androgenized girls is the result of a rather atypical childhood. Perhaps these children were not subjected to quite the same training for "feminine" behavior as are most female children in America. Their parents knew of their history, and might have been more permissive than are most parents with respect to "unfeminine" behavior. Or, conceivably, they might have been *less* permissive with respect to such behavior, thus causing a defensive reaction against "femininity." It is also possible that the girls may have been reacting to what they perceived as unusual in their own situation.[29] Indeed, we cannot even be certain that the girls' behavior really was particularly "tomboyish." Most of the observations were made by individuals who were aware that the girls' sexual physiology was abnormal. Thus, they may have been unusually sensitive to supposed signs of masculinity.

In one of her more recent reports, Ehrhardt indicates that she is aware of these alternative explanations. She argues that since neither the parents nor the girls themselves attach much significance to the genital abnormalities which were present in the girls at birth, it is unlikely that either their behavior or their observations were much influenced by their knowledge of these abnormalities.[30] But this response is not entirely adequate, since such self-reports may be affected by self-deception, or a desire to conform to the interviewer's expectations.

Another possible explanation of the Ehrhardt findings is that prenatal androgenization may increase the level of energy expenditure, without having any direct effect on more specific behaviors, such as aggression. Lenore Tiefer points out that if this hypothesis were true, then

> the androgenital girls would likely ally themselves with other children with the same high energy levels and identify with their (culturally imposed boyish) norms about adult careers, "boys" toys, and clothing preferences.[31]

Tiefer further notes that the lack of any significant differences between the AGS boys and their unaffected brothers, other than a somewhat greater energy expenditure, lends support to this third alternative explanation.

Thus, the evidence for the conclusion that pre- or postnatal exposure to testosterone predisposes human males towards aggressive behavior is far from conclusive. Nevertheless, it would be unwise to dismiss this evidence entirely. That the hypothesis fits so well the stereotype of the naturally aggressive male is grounds for suspicion that the scientists' findings may be more a reflection of their expectations than of reality. Yet it may also show that there is some truth in that stereotype.

Furthermore, even if male hormones are irrelevant to the aggressive behavior of human beings, there may be still be good reasons for concern that increased sex ratios may cause increased violence. If boys learn aggressive behavior from their male peers and elders, then higher sex ratios may make it all the more likely that they will continue to do so. And, given that men often fight about women, a shortage of women may give them more to fight about. Nevertheless, even a preponderantly male population could turn away from war and individual violence. Many individual men and groups of men have done so, in spite of their hormones. If male violence is not an inevitable result of male biology, then altering the relative number of males may have only an indirect effect on its occurrence.

But before we draw this conclusion we need to consider another set of arguments which purport to show that human males have a natural propensity for violence.

Sociobiological Arguments

It was Darwin who first pointed out that in most animal species males and females are subject to somewhat different evolutionary pressures. In most species, males compete with one another for sexual access to females, while females do not compete with one another for sexual access to males. This probably explains why it is usually males which develop more marked secondary sexual characteristics, such as special colors or markings which attract the attention of females, or special weapons such as spurs or antlers, which are used in competition with other males.[32] It probably also explains why males of most species fight among themselves more than do females.

Darwin could not explain why it is usually males who compete for sexual access to females, except by referring to the "stronger passions" of the male—an explanation which verges on circularity. Contemporary sociobiologists have argued that it is the smaller per-offspring investment of males which leads to more intense competition between males. E.O. Wilson notes that, because males usually invest less time and energy in each of their offspring than do females, the potential reproductive success of males is much greater. He believes that this is why there is greater competition among males for sexual access to females.[33] Wilson points out that in the few animal species in which it is males who make the larger per-offspring investment, it is females who compete with one another for sexual access to males. Monogamy evolves only when the reproductive advantage to both mates of rearing their young cooperatively out-weighs the advantage to either of seeking additional mates. If the mating patterns of prehistoric humans were similar to those of most contemporary nonhuman primates, they were nøt monogamous. Thus, it is concluded, they were probably subject to a selection process which encouraged aggression among males more than among females.

Some sociobiologists argue that the hunting adaptation was an additional factor in the evolution of male aggressiveness. At some point in the past, perhaps one to three million years ago, our ancestors began to subsist not only on plant foods, but also on meat. At first, they probably scavenged from the kills of natural predators. Later they learned to make weapons, and to hunt. Among contemporary gathering-and-hunting peoples, large-game hunting is primarily an activity of males. It is possible that this was also the case in the distant past. Adult women would often have been either pregnant or caring for infants, and would thus have been hampered on extended hunting expeditions. Like women in contemporary gathering-and-hunting societies, they may have specialized in collecting and preparing plant foods, along with some small-game hunting.[34] If hunting is an activity requiring more aggressiveness than gathering food and caring for children, then it might seen reasonable to conclude that this sexual division of labor caused males to be more intensively selected for aggressive propensities.

This line of argument has been used by some antifeminists, who maintain that the natural aggressiveness of the human male makes male domination and female subordination inevitable. There are also some feminist writers who, while they reject that conclusion, agree that hunting increased the natural aggressiveness of the human male. Laurel Holliday, for instance, argues that the transition from scavenging meat to killing animals for food caused males to become innately more aggressive. This transition, Holliday says,

> marks the turning point where the notion of profiting by violence came to our ancestors ... Later they turned, as few animals then or now ever do, to killing within their own species ... the decision to kill damaged all of men's relationships—to themselves, to other animals, other men, women, children, and eventually to the cosmos.[35]

Holliday argues that hunting led to warfare because, "When prehistoric men became big-game hunters they became possessive of the territories where their prey could be found."[36] In this way, "human murder had its source in hunting," and eventually became "entrenched in masculine psychology."[37]

If it were true that male biology inevitably drives men towards violent aggression, then the prospect of increased sex ratios due to sex selection would be an extremely disturbing one. We might even be forced to conclude that humanity cannot be turned from its present disastrous course except by using sex selection to greatly reduce the relative number of males. But the evidence does not support the view that human males have a powerful innate drive towards violent behavior. As we have seen, endocrinological studies of both humans and animals suggest that fetal androgenization of the brain and later testosterone production have at most a permissive rather than a deterministic effect upon aggression. Testosterone may affect the ease with which aggressive behavior can be learned, but it does not ensure that it *will* be learned.

Human history presents a depressing spectacle of inhumane behavior on the part of men: "war increases in intensity, bloodiness and duration...through the evolution of culture, reaching its culmination in modern civilization."[38] But it is a mistake to explain this history of male violence primarily in terms of male psychobiology. The evidence of paleontology and anthropology suggests that both organized warfare and high levels of intragroup violence are relatively recent phenomena in human evolution. They appear to be due less to the practice of hunting animals for food than to social, economic and demographic changes which tend to coincide with the *end* of the gathering-and-hunting way of life.

If it were true that the gathering-and-hunting mode of subsistence caused men to be intensively selected for aggressiveness, then we would expect to find that the men in gathering-and-hunting societies are at least as aggressive as those in agricultural or industrialized societies. But the data of anthropology do not confirm this expectation. Scattered over North and South America, central and southern Africa, India, Indonesia, Australia, and Siberia, are societies which have retained the gathering-and-hunting way of life into the modern era. Very few of these societies provide any confirmation of the view that men are by nature highly aggressive.

It is not true that the practice of hunting normally leads to intense competition for territory and warfare between neighboring groups. There was probably no need for such competition during most of human prehistory, because human populations were small and widely distributed. As Ashley Montagu points out,

> There were no hostile neighboring hordes ... for the simple reason that food-gathering hunting populations are always very small in number, scarcely ever exceeding more than a few families living together ... Neighboring populations in prehistoric times would have been few and far between, and when they met it is no more likely that they greeted each other with hostility than do gatherer-hunter peoples today.[39]

Very few gathering-and-hunting societies engage in organized warfare. The aboriginal people of Australia originally constituted several hundred distinct tribes, all living by gathering and hunting. Most observers agree that they were (and, for the most part, still are) a cooperative, friendly, and unaggressive group of peoples. Disputes occurred within and between groups, but were generally settled by arbitration. There was no competition between tribes for territory or conquest, and "no such thing as one tribe taking the field as a whole against another."[40] Disputes within the group sometimes led to shouting, scuffling and the hurling of weapons; but serious injuries were uncommon.[41] Sometimes disputes between groups were settled by ceremonial combat, in which the goal was not to inflict lethal injury (though this sometimes occured), but to resolve the conflict and avoid long-term animosity.[42] Such ceremonial combat was not a form of warfare, but rather "an aspect of juridicial procedure."[43] Where juridicial procedure failed, small groups of men might take vengeance upon another tribe, killing some of the men or capturing women. The women were then legally married to men of the attacking tribe, according to the complex rules of allowable marriage.[44] Such incidents were not common, and never led to large-scale warfare.

The Eskimo peoples of Greenland, northern Canada, Alaska, and Siberia subsisted until recently almost entirely by fishing and hunting. Thus, if hunting were a cause of aggressive behavior, then the Eskimos ought to be unusually aggressive. Yet nothing could be farther from the truth. Birket-Smith, whose observations are not entirely free of ethnocentric bias, nevertheless has only praise for the "gregarious instinct" of the Eskimos. He describes them as generous, friendly, sympathetic, cooperative and unaggressive almost to a fault: "nothing is considered more repulsive than aggressiveness and violence, while on the other hand far-reaching helpfulness among campfellows is an inevitable duty."[45] Generosity and cooperation are necessities in the harsh environment which the Eskimos inhabit. When one hunter succeeds another may fail, but no one goes hungry so long as there is food in the encampment.

The Eskimos did not engage in organized warfare, but murders sometimes occurred. The result might be a long-lasting feud between families, a circumstance which "so disrupted the normal flow of life that an individual might come before his enemies, often unarmed, saying 'kill me!' If they did so, the feud naturally continued. If they did not, however, it might end the matter."[46]

Intrafamilial bloodshed among the Eskimos was quite rare, and so severely disapproved that a particularly belligerent person might be driven away—a fate which meant probable starvation.[47] The point is not that the Eskimos were never aggressive, but that serious violence was generally avoided.

It is well known that the Eskimos practiced infanticide, and that aged, sick or infirm persons were sometimes left to die. This occurred more often among nomadic inland groups than in the coastal settlements, where conditions were less severe. In times of extreme hardship small children and older persons who could no longer be productive were sometimes abandoned, in order that some members of the group might survive. This practice was by not a sign of hostility or aggressiveness towards children or the aged. Children were well treated, and older people had an important role in decision making. The decision to abandon a member of the group was taken only in the direst circumstances, and was made by the group as a whole after careful consideration of all the options.[48]

The Mbuti Pygmies of the Ituri Forest of central Africa are gatherer-hunters whose existence is far less arduous than that of the Eskimos. The rainforest has been their home for thousands of years. It supplies a rich and varied diet, and they regard it as a benevolent deity. But here too, "survival can be achieved only by the closest co-operation and by an elaborate system of reciprocal obligations which ensures that everyone has some share in the day's catch."[49] There are no chiefs or lawgivers, and a good deal of bickering and minor mayhem occurs. Yet disputes are nearly always settled without serious violence. Minor wrongdoing is punished with barbed humor or temporary ostracism. Major wrongdoing, such as murder, is never mentioned in Colin Turnbull's classic account of the years which he spent with the Pygmies. Turnbull reports that warfare and feuding are nonexistent.[50] When disputes occur,

> Settlement is reached with one goal in mind, and that is the restoration of harmony within the band, for the good of the whole. If there is one thing that is surely wrong in their eyes, it is that the dispute should have taken place to begin with, and to this extent both disputants are to blame and are held in temporary disfavour.[51]

The story is much the same in other gathering-and-hunting societies, at least where traditional ways of life have not been too badly disrupted. So long as people live in small groups, subsisting on foods provided by the natural environment, there is little to be gained and much to be lost from the toleration of violent behavior. Consequently, even though males may tend to be more violent than females, severe violence is kept to a minimum. What violence there is tends to occur in small isolated incidents, rather than wholesale outbreaks.

There is, in fact, no conclusive evidence of large-scale violence between human beings prior to the agricultural revolution. Most of the evidence which has been presented as proof of the ferocity of our earlier ancestors is susceptible to other interpretations. Raymond Dart, for instance, interpreted damage to certain of the australopithecine skulls found in South African cave deposits as due to lethal blows which could only have been inflicted by club-wielding hominids.[52] But more recent work has shown that the damage could just as well have been inflicted long after death, e.g., by objects resting above the skulls and gradually pressed into them by the weight of material deposited above.[53]

Other fossil bones, including homo erectus skulls found at Choukoutien, China, and Neanderthal skulls found at various sites in

Europe, show apparent evidence of cannibalism. In some, the lower surface of the brain case is missing, and appears to have been removed in order to extract the brain. Neanderthal bones have been found which bear marks of fire and cutting. These have been interpreted as the remains of cannibal feasts.[54] However, it is difficult to determine whether these individuals were killed in battle or slaughtered for food, or whether the damage is the result of funerary practices. Cremation, decarnization, and cannibalism are sometimes part of traditional funerary customs. Thus, evidence of such practices does not prove that prehistoric humans waged war, or that they frequently murdered one another.

Male belligerence is not advantageous to a gathering-and-hunting society, for it endangers the cooperation upon which survival depends. In such societies, men usually cannot improve their social or material status through violence against other people. The successful hunter usually does not achieve social dominance either. Among the !Kung, for instance,

> The stress on equality demands that certain rituals are observed when a successful hunter returns to camp. The object of these rituals is to play down the event so as to discourage arrogance and conceit ... A !Kung man ... described it this way: 'Say that a man has been hunting. He must not come home and announce like a braggart, "I have killed a big one in the bush!" He must first sit down in silence until I or someone else comes up to his fire and asks, "What did you see today?" He replies quietly, "Ah, I'm no good for hunting. I saw nothing at all ... maybe just a tiny one." Then I smile to myself because I now know he has killed something big.' The bigger the kill, the more it is played down ... [This custom is followed] not just by the !Kung but by many foraging people, and the result is that although some men are undoubtedly more proficient hunters than others, no one accrues unusual prestige or status because of his talents.[55]

Thus, there is little reason to believe that, during the millions of years in which our ancestors lived by gathering and hunting, males were intensively selected for aggressive behavior. If gathering-and-hunting peoples usually manage to conduct their affairs without much serious violence, then we must conclude that the violence of today's world is more a function of social and economic conditions than of male psychobiology. Our ancestors abandoned the gathering-and-hunting way of life much too recently for the fundamentals of our genetic inheritance to have been greatly altered since that time. Thus, male hormones may help to explain why it is most often males who learn to get what they want through aggression, but they do not explain why it is necessary or profitable for *anyone* to learn the lessons of violence. For that we have to look to the conditions of scarcity, competition, exploitation, and oppression which lead individuals, groups and nations to conclude that the use of force is the best means of achieving their ends.

Sex Ratios and Levels of Violence

If the violent behavior of men is more a function of social conditions than of male psychobiology, then there is probably no general relationship between sex ratios and the amount of violence in a society. An increase in the relative number of males could coincide with other changes which lead to a reduction in the amount of violence—e.g., decreases in population density, or a juster distribution of resources. Conversely, a relative increase in the number of females might do nothing to decrease the occurrence of violence, if it were unaccompanied by such other changes.

In the United States, it is not those ethnic groups having the highest sex ratios which have the highest levels of violent crime, but rather those which are subject to the most severe economic deprivation. The President's Commission on Law Enforcement and the Administration of Justice studied the sociological correlates of violent crime in American urban areas, and listed the following as particularly significant: high unemployment, low income, high population density, substandard housing, poor health and medical care, poor education, racial prejudice, "broken homes," and alcohol and drug abuse.[56] Ironically, these conditions tend to coincide with unusually low sex ratios, especially among urban black Americans. According to Guttentag and Secord,

> in 1970, for the age cohorts most eligible for marriage, there were almost two black women for every man. This is for the United States as a whole. Regional variations create even worse situations in some locations, and better ones in others.[57]

The reasons for this sexual imbalance include the high number of black men in the armed forces and penal institutions; low sex ratios at birth due to higher fetal death rates for males; higher mortality rates for male infants and black males throughout the life span; and the migration of larger numbers of black women to northern urban areas.[58] Obviously, none of these conditions have anything to do with race as such. The point is that in this context it would be difficult to argue that low sex ratios are conducive to a nonviolent society. In a racist and sexist society, where it is extremely difficult for unmarried women to adequately support their families, a scarcity of males can aggravate the poverty and deprivation which contribute to violence.

It is also possible to cite instances in which high sex ratios correspond with a high incidence of violence. The frontier societies of America and Australia were noted both for their violence and for the scarcity of white women. Many highly militaristic societies—such as the Caribs,[59] the Yanomamo of Brazil,[60] the Rajputs of northern India,[61] and certain Arab tribes[62]—have had high sex ratios due to the extensive practice of female infanticide. Yet there are other high-sex-ratio societies which are noted for their peaceful ways. The Eskimos exposed female infants more often than males, and consequently suffered from a chronic shortage of women.[63] Yet this shortage did not lead to warfare, or to a high level of violence within social groups. The gentle Arapesh raise more boys than girls, while the ungentle Mundugumor raise more girls than boys.[64]

Thus, high sex ratios do not automatically produce a more violent culture. Where there is an association between the two phenomena, it is at least as likely that the violence of the society contributes to son-preference and the consequent high sex ratios than the reverse. The American and Australian frontiers were probably not violent because of the scarcity of women. Rather, women were scarce because life on the frontier was difficult and unsafe. The violence must be traced to other causes—such as the habits and ideology which the white men developed or brought with them.

Because there is no regular correlation between high sex ratios and high levels of violence, it is difficult to predict whether increased sex ratios will contribute to increased violence. It is probable that in most cases sex-ratio changes will be far less important than other social and environmental factors. Of these, none is more important than radical inequality in the distribution of wealth, both between and within nations. In the United States, the median income of blacks is still only 55 percent that of whites, and twice as many blacks are unemployed. The global situation is even worse:

> A quarter of the world's population lives in the "developed" nations of the northern hemisphere, and they account for four-fifths of the global income. The poorer countries ... are inhabited by the remaining three-quarters of the population who share just one-fifth of the world's wealth.[65]

So long as such gross inequalities exist, the prospects for world peace and social stability within nations will remain dim, whatever the relative number of women and men.

Social injustice also contributes to many physiological factors which are believed to be related to violent behavior. Poor prenatal diet increases the risk of toxemia during pregnancy and of premature birth, both of which are thought to contribute to hyperactivity, aggressiveness and antisocial behavior.[66] This is especially true of male infants, who seem to be more vulnerable to prenatal brain damage. Poor postnatal nutrition can also retard or permanently impair the development of the central nervous system, which in turn may contribute to aggressiveness, and even psychopathology. Behavioral disorders have been linked with shortages of a wide range of specific nutrients, as well as to generally poor nutrition.[67] The best nutrition will not prevent the occurrence of violence where people suffer from a perception of injustice. Nevertheless, the efficiency with which people's brains function, and their readiness to resort to violence, may be affected by how well they have been nourished.

These theories about the physiological causes of violence are somewhat speculative. Much more thoroughly established is the connection between the violent punishment of children and later violent behavior. It is consistently found that boys are more apt to be physically punished than girls, and that the more violently a child is punished the more likely s/he will becme a violent adolescent or adult.[68] There is a problem about the direction of the causal relationship here, since it is probable that aggressive children are more apt to be physically punished. Yet it seems clear that the more violence a child experiences, the more violent s/he is likely to become.

None of this means that increased sex ratios might not also contribute to violence. There is enough evidence that the greater propensity of males towards violent behavior is in some part due to male biology to suggest that parents who opt for daughters because they do not want to raise potential rapists or batterers are not acting irrationally. Yet it is impossible to know whether the use of sex selection to produce sons will contribute significantly to a more violent world. If it does, it will not be because high sex ratios by themselves necessarily increase the incidence of violence, but rather because of complex interactions between sex ratios and many other social variables. Thus, very little weight can be placed on the argument that the use of sex selection is immoral because higher sex ratios will mean more violence.

Notes

1. *Violent Crime*, Report of the National Commission on the Causes and Prevention of Violence (New York: George Braziller, 1969); cited by Kenneth E. Moyer, "Sex Differences in Aggression," in *Sex Differences in Behavior*, edited by Richard C. Friedman, Ralph M. Richart, and Raymond L. Vande Wiele (Huntington, New York: Robert E. Krieger Publishing Company, 1978), 335.

2. Roy G. D'Andrade, "Sex Differences and Cultural Institutions," *The Development of Sex Differences*, edited by Eleanor E. Maccoby (London: Tavistock Publications, 1967), 201.

3. Eleanor Emmons Maccoby and Carol Nagy Jacklin, *The Psychology of Sex Differences* (Stanford California: Stanford University Press, 1974), 351–52.

4. Amitai Etzioni, "Sex Control, Science, and Society", *Science* 161 (September 13, 1968), 1109.

5. Edward C. Whitmont, *Return of the Goddess: Femininity, Aggression and the Modern Grail Quest* (London: Routledge and Kegan Paul, 1982), viii.

6. Whitmont, 12.

7. *Webster's New School and Office Dictionary* (Greenwich, Connecticut: Fawcett Publications, 1960), 91–94.

8. Steven Goldberg, *The Inevitability of Patriarchy* (New York: William Morrow and Company, 1973/74), 91–94.

9. John R. Lion and Manoel Peuna, "The Study of Human Aggression," in *The Neurophysiology of Aggression*, edited by Richard E. Whalen (New York: Plenum Press, 1974), 167.

10. Margaret Mead, *Sex and Temperament in Three Primitive Societies* (New York: Morrow Quill Paperbacks, 1963), 15.

11. Mead, 24–25.

12. Mead, 208–9.

13. Mead, 50.

14. F.H. Bronson and C. Desjardins, "Steroid Hormones and Aggressive Behavior in Mammals," *The Physiology of Aggression and Defeat*, edited by Basil E. Eleftherion and John Paul Scott (New York: Plenum Press, 1971), 44–46.

15. Charles H. Phoenix, "Prenatal Testosterone in the Nonhuman Primate and Its Consequences for Behavior," in *Sex Differences in Behavior*, 19–32.

16. Bronson and Desjardins, 43.

17. See, for instance, Robert Ardrey, *African Genesis* (New York: Atheneum, 1963), and *The Territorial Imperative* (New York: Atheneum, 1966); and Lionel Tiger, *Men in Groups* (New York: Random House, 1969).

18. David Pilbeam, "An Idea We Could Live Without: The Naked Ape," *Man and Aggression*, edited by Ashley Montagu (New York: Oxford University Press, 1973), 112–15.

19. Bronson and Desjardins, 52.

20. John Stuart Mill, *The Subjection of Women*, in *Essays on Sex Equality*, by John Stuart Mill and Harriet Taylor Mill, edited by Alice S. Rossi (Chicago: University of Chicago Press, 1970), 148.

21. John Money and Anke Ehrhardt, *Man and Woman, Boy and Girl: The Differentiation and Dimorphism of Gender Identity from Conception to Maturity* (New York: Mentor Books, 1974).

22. Anke Ehrhardt and Susan W. Baker, "Fetal Androgens, Human Central Nervous System Differentiation, and Behavioral Differences," in *Sex Differences in Behavior*, 32–52.

23. Anke Ehrhardt, "Maternalism in Fetal Hormonal and Related Syndromes," in *Contemporary Sexual Behavior: Critical Issues in the 1970's*, edited by Joseph Zubin and John Money (Baltimore: Johns Hopkins University Press, 1973), 99–116.

24. Money and Ehrhardt, op. cit.

25. Ibid.

26. H., Persky, K.D., Smith, and G.K., Basu, "Relation of Psychologic Measures of Aggression and Hostility to Testosterone Production in Men," *Psychosomatic Medicine* 33 (1971), 265–77.

27. Leo E. Kreuz and Robert M. Rose, "Assessment of Aggressive Behavior and Plasma Testosterone in a Young Criminal Population," *Physiology of Aggression and Implications for Control*, edited by Kenneth Evan Moyer (New York: Raven Press, 1976), 219–30.

28. Kreuz and Rose, 228.

29. Helen Longino and Ruth Doell, "Body, Bias and Behavior: A Comparative Analysis of Reasoning in Two Areas of Biological Science," *Signs: A Journal of Women in Culture and Society* 9:2 (Winter 1983), 220.

30. Ehrhardt and Baker, 49.

31. Leonore Tiefer, "The Context and Consequences of Contemporary Sex Research: A Feminist Perspective," in *Sex and Behavior, Status and Prospects*, edited by Thomas E. McGill, Donald Dewsbury and Benjamin D. Sacks (New York: Plenum Press, 1978), 374.

32. Charles Darwin, *The Descent of Man and Selection in Relation to Sex* (New York: D. Appleton, 1874).

33. Edward O. Wilson, *Sociobiology: The New Synthesis* (Cambridge, Massachusetts: Harvard University Press, 1975).

34. Except in the coldest climates, where plant foods are unavailable during much of the year, women's gathering activities usually provide the largest part of the diet of gathering-and-hunting people. This is why it is better to speak of "gatherer-hunters," rather than "hunter-gatherers."

35. Laurel Holliday, *The Violent Sex: Male Psychobiology and the Evolution of Consciousness* (Guerneville, California: Bluestocking Books, 1978), 113–114.

36. Holliday, 118.

37. Holliday, 120.

38. Marshall Sahlins, quoted by Richard E. Leakey, *The Making of Mankind* (London: Michael Joseph, 1981), 219.

39. Ashley Montagu, "Aggression and the Evolution of Man," *The Neurophysiology of Aggression*, 2.

40. Gerald C., Wheeler, *The Tribe, and Intertribal Relations in Australia* (London: John Murray, 1910), 155. Also see Elman R. Service, *A Profile of Primitive Culture* (New York: Harper and Brothers, 1958), 18.

41. A.P. Elkin, *The Australian Aborigines* (Sydney: Angus and Robertson, 1938), 200, 203; Wheeler, 148–54.

42. Harney, W. E., *Taboo* (Sydney: Australian Publishing Company, 1943).

43. Service, 18.

44. Wheeler, 154.

45. Kaj Birket-Smith, *The Eskimos* (London: Methuen and Company, 1959), 52–54.

46. Robert F. Spencer, *The North Alaskan Eskimo: A Study in Ecology and Society* (Washington D.C.: United States Government Printing Office, 1959), 72.

47. Spencer, 73.

48. Spencer, 92–95.

49. Colin M. Turnbull, *The Forest People* (New York: Simon and Schuster, 1961), 156.

50. Turnbull, Discussion, in *Man the Hunter*, edited by R.B. Lee and I. Devore (Chicago: Aldine, 1968).

51. Turnbull, *The Mountain People* (London: Jonathan Cape, 1973), 156.

52. Raymond Dart, "The Osteodontokeratic Culture of Australopithecus Prometheus," *Transvaal Museum Memoir* 10 (1957).

53. Leaky, 224.

54. Henry V. Vallois, "The Social Life of Early Man: The Evidence of Skeletons," in *Social Life of Early Man*, edited by Sherwood L. Washburn (London: Methuen and Company, 1962), 231–32.

55. Leaky, 107.

56. President's Commission on Law Enforcement and Administration of Justice, *Challenge of Crime* (Washington, D.C.: U.S. Government Printing Office, 1967).

57. Marcia Guttentag and Paul Secord, *Too Many Women? The Sex Ratio Question* (Beverly Hills: Sage Publications, 1984), 201.

58. Guttentag and Secord, 207–14.

59. John F. McLennan, *Primitive Marriage* (Chicago: University of Chicago Press, 1970), 27–28 and 37–38.

60. Napoleon A. Chagnon, *Yanomamo: The Fierce People* (New York: Holt, Rinehart and Winston, 1968), 74–75.

61. William Crook, *Religion and Folklore of Northern India* (New Delhi: S. Chand and Company, 1925), 144.

62. W. Robertson Smith, *Kinship and Marriage in Early Arabia*, (The Netherlands: Anthropological Publications, Oosterhout N.B. 1966), 153.

63. See Margaret Mead, *Male and Female* (New York: Dell Publishing Company, 1971), 203; also Birket-Smith, 139, and Spencer, 93–94.

64. Mead, *Sex and Temperament*.

65. Leakey, 243.

66. See, for instance, G. J. Mohr and P.F. Bartelme, "Mental and Physical Development of Children Prematurely Born," *American Journal of Diseases of Children* 40 (1930); B. Pasamanick et al., "Pregnancy Experience and Development of Behavioral Disorder in Children," *American Journal of Psychiatry* 112 (1956), 613; Mary Shirley, "A Behavioral Syndrome Characterizing Prematurely Born Children," *Child Development* 10 (1939), 115; and A. Zitrin et al., "Pre- and Paranatal Factors in Mental Disorders of Children," *Journal of Nervous and Mental Disorders* 139 (1964), 357.

67. See Roger J. Williams, *Nutrition Against Disease* (New York: Bantam Books), 1971, 151–69; and Linus Pauling, "Orthomolecular Psychiatry," *Science* April 19, 1968, 265–71.

68. Eleanor Maccoby and Carol Jacklin, *The Psychology of Sex Differences* (Stanford, California: Stanford University Press, 1974).

6

Other Consequentialist Objections

There are many other grounds for concern that the widespread practice of sex selection may be detrimental. The possible ill effects include: (1) stricter confinement of women to subordinate "feminine" roles; (2) increased numbers of male firstborns, with consequent psychological damage to women; (3) increased dependence of women upon the male-dominated medical profession; (4) unhappiness among "excess" men unable to find wives; (5) increased prostitution; (6) continued destruction of the natural world due to the increased hegemony of masculinist attitudes; (7) intensified class conflict due to unequal sex ratio changes in different social groups; and (8) a slide towards more dangerous forms of human genetic engineering. While none of these results is certain, each is indeed possible.

Sex Ratios and the Status of Women

What will happen to women's rights and women's social status if sex ratios increase due to sex selection? Very few studies have been made of the relationship between sex ratios and the social status of women. The only well-developed theory in this area is that presented by Marcia Guttentag and Paul Secord.[1] In their view, high-sex-ratio societies tend to be severely patriarchal, imposing rigid restrictions upon women's sexual behavior and confining them to the domestic role. "Good" women are respected, but kept firmly in their place. In low-sex-ratio societies, the restrictions upon women's sexual behavior are less stringent, and feminist movements are more likely to arise.

However, Guttentag and Secord do not regard low sex ratios as altogether beneficial to women. On the contrary, they argue that low sex ratios contribute to male misogyny, the devaluation of women and marriage, and a subjective sense of powerlessness on the part of women. This is because the sex which is in the minority tends to have greater "dyadic" power, i.e., greater power in two-party heterosexual relationships.

> The individual ... whose sex is in short supply has a stronger position and is less dependent on the partner because of the larger number of alternative relationships available to him or her [and thus is] able to negotiate more favorable outcomes within the *dyad*, or two-person relationship.[2]

Where women are scarce (so the argument goes), young adult women are highly valued, and most men want to marry and are willing to make a long-term commitment to one woman. Although generally confined to the subordinate domestic role, women may enjoy a subjective sense of power, particularly if they are able to choose their husband from among competing suitors. They may gain economic mobility by marrying upward in social class, whereas many men will be unable to afford marriage. Female virginity will be prized and women will be severely punished for extramarital sexual activities. Yet in spite of their lack of freedom, most women will be satisfied with their role. They will be unlikely to have strong career aspirations or to demand stronger legal rights or greater personal freedom.

On the other hand (the argument continues), where there is an oversupply of marriageable women, women are less highly valued and have a subjective sense of powerlessness. Men have more opportunities for sexual relationships with women, and thus are less willing to make long-term commitments to individual women. Women have less chance of upward mobility through marriage and are more apt to find themselves divorced, abandoned, or forced to share a man with other women. Unless polygyny is practiced, many women will be single during all or part of their adult lives. Sexual restrictions will be relaxed, but this relaxation may benefit men more than women, who are apt to be exploited as sex objects. Under such conditions, some women will redouble their efforts to be pleasing to men, while others will become resentful of male domination and seek greater economic independence and increased political influence. Many of these characteristics of the typical low-sex-ratio society may be observed in contemporary industrialized nations where women are in the majority.

Guttentag and Secord find support for this account of typical high- and low-sex-ratio societies in past societies such as Athens and Sparta. Athens was a typical high-sex-ratio society. "Respectable" women were strictly confined and subjected to an exceptionally severe double standard of sexual morality. Sparta, with much lower sex ratios, allowed women more sexual freedom, more freedom of movement, and more extensive legal rights.[3]

Although these contrasting tendencies are typical of high- versus low-sex-ratio societies, they vary considerably, according to the extent of male structural power. Structural power is defined as control over the legal, political, economic, and other key institutions of the society. Because males usually have the monopoly of structural power, they are often able to limit women's freedom even when sex ratios are high. Indeed,

> Where men hold the balance of political and economic power, excluding women from direct participation, high sex ratios often make matters worse. Women are apt to be controlled by men and to have few rights of their own.[4]

Guttentag and Secord point to the situation of Chinese-American women in the nineteenth century as one example of this sort. Chinese men outnumbered Chinese women in America by as much as 20 to 1, and held control of the tongs and other traditional social structures. As a result, many Chinese women were imprisoned in brothels, and those who were married were often kept as virtual prisoners.[5] A similar fate befell many of the women deported to Australia in the early decades after the founding of the first English penal colony in 1788. Vastly outnumbered and powerless, they were forced into prostitution or concubinage.[6]

On the other hand, where men have less structural power they may be unable to limit women's freedom even when sex ratios are high. Guttentag and Secord cite as an example the Bakweri people in the plantation economy of Cameroon, West Africa during the 1950s. Because of the in-migration of men, the sex ratio was over 2 to 1 in favor of males. Yet women had a great deal of freedom to change husbands or carry on extramarital affairs, through which they often gained a more substantial income than was possible for most men. In this case, men's lack of economic power made it impossible for them to impose strict controls upon women.[7]

On this theory, high sex ratios may be either good or bad for women, depending upon the intensity of the male monopoly of structural power. If women are economically dependent and lack fundamental legal rights, they are not likely to gain much advantage from their own scarcity. But if they have a degree of economic independence and legal autonomy, then they may be able to use their increased dyadic power to drive better bargains in sexual relationships with men. Of course, this benefit applies primarily to heterosexual women. Guttentag and Secord say very little about nonheterosexual women, except that lesbianism is apt to be more severely discouraged in a high-sex-ratio society. Since compulsory heterosexuality is a basic part of the oppression of women, this must be seen as another possible detrimental effect of high sex ratios.

Low sex ratios may also prove either good or bad for women, depending upon the intensity of male structural power. Where polygyny is not practiced and single women have few means of earning an adequate living, a shortage of men is bound to result in more women living in poverty. In contrast, women who have adequate means of survival outside of marriage may benefit from low sex ratios. The more independent women there are, the greater their potential political influence. Moreover, where men are scarce, tasks that would previously have been done by men must be done by women. Since "men's work" is usually accorded greater respect than "women's work," women may profit from this change. During the Second World War, American women moved into many of the jobs vacated by men. Although most lost their jobs when the soldiers returned, the experience of economic independence was not forgotten. The higher casualties suffered by the Soviet Union created a longer-lasting shortage of men, and there women have continued to perform many jobs which in the United States are done mostly by men.

Thus, the better women's status to begin with, the more they stand to gain from either an increase or a decrease in sex ratios. The worse their status to begin with, the more they are apt to lose in either case. However, if Guttentag and Secord are right, the gains and losses tend to be different in the two cases. The potential benefits to women of increased sex ratios include greater opportunities to form stable relationships with men, and a greater subjective sense of power within the traditional female roles. The potential liabilities include a loss of freedom to deviate from these traditional roles, exclusion from most high-status positions, and the absence of a strong feminist movement to combat such injustices. The potential benefits of decreased sex ratios include greater sexual freedom, expanded extradomestic economic opportunities, and a stronger feminist movement. The liabilities include greater difficulty in forming stable relationships with men, and more misogyny and sexually exploitative behavior on the part of men.

But even if Guttentag and Secord have correctly described the properties of typical high- and low-sex-ratio societies, their theory may not generate any reliable predictions about the results of sex-ratio increases in the future. For the differences between typical high- and low-sex-ratio societies may not be caused by sex-ratio differences. It may be that there is no general causal relationship between sex ratios and the typical properties of high- and low-sex-ratio societies, or that the causal relationship tends to run in the opposite direction. For instance, rather than high sex ratios causing women to be confined to the domestic role, it may be that societies which confine women to this role tend to have high sex ratios because parents conclude that raising females is less worthwhile than raising males, and therefore practice sex selection through female infanticide or the neglect of female children. If it is true that when sex ratios are low men tend to become more misogynous, this may be less because low sex ratios produce misogyny than because low sex ratios are often the result of social disruptions such as war or plague. These disruptions may be the underlying cause of the resentment which, for cultural reasons, is expressed as misogyny.

Furthermore, even if high and low sex ratios in the past have had the effects which Guttentag and Secord describe, we cannot assume that they will have similar effects in the future. We may speculate that women in the more severely patriarchal societies will suffer a further loss of freedom should sex selection lead to an increase in sex ratios. However, it is also possible that industrialization, improved education, increased availability of contraception and abortion, movements towards socialism and democracy, and other factors will loosen traditional constraints on women, tipping the balance towards sexual equality despite increased sex ratios. The technologies of mass communication will make it more difficult for women to be kept in a state of ignorance. Women are gaining access to these modes of communication, and forming national and international networks of feminist activists and researchers. Thus, it will be more difficult to suppress feminist awareness or conceal feminist history. It is also possible that if sex ratios begin to increase precipitously due to sex selection, governments will see the need to improve women's legal and economic situation. How else can parents be persuaded that it is in their interest to raise daughters? Some reports from China indicate that a dynamic of this sort may be beginning to operate there.

Women in the less severely patriarchal nations probably have less to fear from sex-ratio increases due to sex selection. Not only are sex-ratio increases apt to be smaller but, because women have greater opportunities for economic independence and political influence, they may be able to use their increased dyadic power to improve their legal and economic status. Older women may find it easier to find male partners—not a trivial consideration, given that a majority of married women face a period of widowhood, and that many divorced or widowed women who would like to remarry cannot. Women may find it easier to leave abusive men when more alternative relationships are possible. To the extent that many jobs remain sex-segregated, a shortage of women may make it easier for women in predominantly female jobs to demand more adequate wages and working conditions. Of course, none of this is predetermined either. The forces of reaction are strong, and the gains which women have made may still be lost, with or without sex-ratio increases.

Thus, sweeping predictions of a loss of freedom for women should sex ratios increase are unjustified. Women in each society, each economic class and ethnic group will have to study their own situation to arrive at informed judgments about how they and their successors may be affected by sex-ratio changes.

It is also worth pointing out that women differ in their attitudes towards the hypothetical advantages and disadvantages of high- and low-sex-ratio societies. Women whose aspirations are not confined to wife and motherhood rightly regard the restrictions of the typical high-sex-ratio society as intolerable. But women who (expect to) find their primary satisfactions in such traditional roles may prefer the secure domestic status which high-sex-ratio societies may offer. Antifeminist women are often those who have decided that, for them, freedom is not worth the liabilities which it may entail.[8] Their preferences must be regarded as authentic, though perhaps unfortunate. Thus, it is impossible to argue that either the typical high-sex-ratio society or the typical low-sex-ratio society is better for *all* women, according to their own evaluation.

The Advantages of the Firstborn Child

A majority of prospective parents throughout most of the world would prefer a male firstborn child. Although many would allow the sex of the first child to be determined by chance, resorting to sex selection only with later children, many others would choose to have a male child first. Thus, were an effective method of selecting sex to become widely available, there would probably by a significant increase in the relative number of male firstborns. This prospect is alarming to those feminists who believe that there are important psychological advantages in being a firstborn child.[9]

The psychological effects of birth order have been studied for over a century. In 1874 Sir Francis Galton published data on eminent English scientists (all of them male), and concluded that firstborn and only children are more apt to be high achievers.[10] Galton realized that much of the overrepresentation of firstborn and only sons among prominent scientists was due to the law of primogeniture, which gave firstborn sons superior inheritance rights and more extensive educational opportunities. But he also argued that firstborn sons are given greater responsibilities than later-born sons, and that as a result they tend to be more achievement oriented.

Dozens of studies have shown that firstborns of both sexes are overrepresented among prominent persons in many professions, among students attending college, and among those scoring highest on aptitude and intelligence tests.[11] In addition, hundreds of studies have appeared to establish or disprove linkages between birth order and such traits as motivation, initiative, creativity, popularity, conformity, anxiety, affiliation, dependence, conservatism, rebelliousness, authoritarianism, mental illness, juvenile delinquency, criminality, alcoholism, and marital adjustment. The results are enormously complex and frequently contradictory. However, among the most consistent findings are that firstborns tend to achieve more in formal education and careers, and that they tend to be more dependent and affiliative.[12]

Much of the interest in the psychological effects of birth order has been inspired by Alfred Adler. Adler maintained that a child's order of birth influences his or her relationships with parents and siblings, and thus has a critical influence upon personality and psychological development.[13] Adlerian theorists argue that each birth-order position carries certain advantages and disadvantages. Firstborns initially enjoy more of their parents' attention, and more is expected of them. Consequently, they are apt to be hard-working high achievers. On the other hand, they experience a traumatic "dethronement" after the arrival of the second child, and may therefore suffer from envy, insecurity, and other psychological problems. Though often conservative and respectful of authority, they may rebel against the heavy demands placed upon them. Middle children have a competitive disadvantage and are less apt to achieve in the traditional ways; but they are also less dependent, socially better adjusted, and more apt to pursue alternative forms of achievement. Youngest children share some of the advantages of firstborns, without the trauma of dethronement; they are usually the last to leave home and thus, for a time, face less domestic competition than do older siblings. Only children are also spared the experience of dethronement; but because they are never forced to compete with a sibling they may be less achievement oriented, or more socially inept. All of these patterns are complicated by the further influences of the sex and temporal spacing of siblings.

Another theory which seeks to account for the greater achievements of firstborns is the "confluence model" developed by Robert Zajonc. On this theory, the oldest child is apt to be the most intelligent, and successive children progressively less intelligent. This is because each additional child produces a degradation of the family's "intellectual environment." A family's intellectual environment is defined as the average of the present intellectual capacities of all its members. Zajonc argues that the intellectual environment deteriorates with each birth simply because children know less than adults, and younger children less than older ones.[14] Older children have an additional advantage over both younger and only children, in that they have more opportunity to teach: "it is better to have a younger sibling you can show the world to than an older one who shows it off to you."[15]

If either of these theories about the psychological effects of birth order were well supported, there might be good reason to fear that increases in the relative number of male firstborns will have a detrimental effect upon women. It is interesting to note, however, that some proponents of the Adlerian theory believe that second-born daughters have certain advantages over firstborn daughters. For instance, it has been suggested that they are more apt to rebel against traditional feminine roles than firstborn daughters, and are more motivated to achieve in the "man's world"—particularly if their older sibling is male.[16] Furthermore, firstborn girls do not share all of the advantages of their male counterparts, because they are not members of the socially dominant gender, and because they may be saddled from an early age with child-care responsibilities.[17]

But we need not pursue the question of whether, on either of these theories about birth-order effects, the use of sex selection to produce male firstborns is likely to be on the whole good or bad for women. For, although both theories have some intuitive plausibility, neither is supported by the more careful research on birth-order effects.

The isolation of birth-order effects from the effects of socioeconomic status, ethnicity, religion, family size, urban versus rural background, demographic change, and other variables represents an extremely difficult methodological problem, and one which has not been resolved in most of the studies that have been done. In many of the early studies which found firstborns to be superior in intelligence, motivation and achievement, there were no controls for family size. Obviously, firstborns are more apt to come from small families. In most of the industrialized nations, parents of large families tend to have less money and education, and to score lower on standard tests of intelligence than parents of small families. Thus, where family size is not held constant, comparisons between first- and later-born children are biased. The latter tend to have less privileged social backgrounds, and any statistical differences in intelligence, motivation or achievement are at least as likely to be due to this factor as to birth order. Where sample groups of first- and laterborns are matched for family size and socioeconomic status, most (though not quite all) of the apparent superiorities of the firstborn disappear.[18]

Demographic change is another factor which has biased some studies of birth-order effects. In 1972, Carmi Schooler showed that much of the overrepresentation of firstborns among gifted students and college populations during the 1960s was an artifact of the postwar baby boom.[19] Because more families were started in the postwar years, firstborns were statistically overrepresented in the high school and college-age populations as a whole, and not only among gifted and college students.

The birth-order debate continues, with some psychologists continuing to argue that birth order has a significant effect upon personality and others debunking the idea. But at present the evidence appears to support a sceptical view of the intrinsic importance of birth order. In 1983 two Swiss psychiatrists, Cécile Ernst and Jules Angst, published an exhaustive review of the birth-order research of the past four decades, comprising some 1,500 studies. They conclude that nearly all of the reports of birth-order effects are due to errors in the design of the studies and the analysis of the data. They argue that,

> higher IQ and greater socialization in first- than in laterborns are generally not differences between sibs, but between subjects in large and small families, between middle- and lower-class subjects, between subjects of different religious denominations and ethnicity, and between persons living in cities and in the country.[20]

These authors present a convincing case that birth order has not been shown to have any consistent effect upon ability or personality development. They believe that there are some general differences in the socialization process undergone by firstborns and later borns. Firstborns tend to be better cared for in infancy and to be more precocious in their linguistic development; and they do experience a "dethronement" with the birth of younger siblings. But these differences "do not seem to leave indelible traces that can be predicted."[21] They do not deny that being a first, middle, last or only child may be important to the personal development of some individuals, but only that it has any general and lasting significance.

Ernst and Angst argue that this conclusion makes sense, because each child is unique and will react to a given situation—such as being a firstborn or only child—in a way different from any other. The Adlerian theory does allow for a variety of development patterns in persons of a given birth order. But this, rather than strengthening the theory, robs it of much of its apparent content. If, for example, being firstborn may make one either a conservative or a rebel, either highly stable or mentally disturbed, then it would seem that virtually any outcome can be "explained" in terms of birth order, but that no particular outcome can ever be predicted.

It seems, then, that the existing theories about the effects of birth order upon personal development are either empirically unsubstantiated or devoid of predictive power. If there are any general birth order effects, they are probably rather small. Moreover, the complexity and inconsistency of the various effects which have been alleged to occur makes it virtually impossible to say whether it is most advantageous to be born first, middle, last, or as an only child. Thus, no strong moral objection to sex selection follows from the simple fact that it is apt to result in a higher proportion of male firstborns.

Other Psychological Effects

But what about the fact that many females will know that their parents chose to have sons first? Roberta Steinbacher asks, "What are the implications of being second born, and knowing at some early age that you were planned-to-be-second?"[22] Will girls suffer a loss of confidence or self-esteem from such knowledge? John Fletcher argues that even a firstborn girl may be damaged if she learns that whereas she was not sex-selected, her younger brother was. Addressing the parent who already has a daughter and is considering a sex-selected son, Fletcher says,

> put yourself in your daughter's place. How will she respond to your reasons *why* you went to the fertility clinic to start a pregnancy with baby brother, when you did not do the same with the conception of her? What reasons will you give her? ... You would not let her continue believing that only boys can be police, firefighters or surgeons, would you? ... You conclude that if you would not neglect her need to aspire equally to almost any job that a man might do, you will sabotage that parental duty by preselecting sex.[23]

This argument rests upon the assumption that there can be no nonsexist reasons for preselecting sex—or none which a female child can be expected to understand. But we have seen that this assumption is false. One need not believe that girls cannot grow up to be firefighters or surgeons to want a son as well as a daughter. One might, for instance, believe that children develop a better understanding of persons of the opposite sex if they have an opposite-sex sibling, Or, one might believe that the best way to raise a nonsexist child is to raise him or her in the company of an opposite-sex sibling whom one does not treat any differently. Even if these beliefs are false, they are not obvious instances of sexism. Nor is it obvious that a girl would be apt to be disturbed by learning that her parents selected a son because of such beliefs.

What of the girl who learns that her brother was sex-selected for reasons which *are* sexist—e.g., because her father wants a male heir? No doubt she may be hurt by this knowledge. Yet, if her parents are sexist in their current behavior, if they treat her as worth less than her brother, then the discovery that his sex was preselected can only come as one more confirmation of what she must already know. On the other hand, if her parents are not sexist in their current behavior, then this discovery need not shake her conviction that she is equally valued—although of course it might. She will, in any case, learn that her society values males more highly than females. This message is conveyed to female children in a multitude of ways. Any resulting injury to female self-esteem is highly overdetermined. Thus, it is not likely that knowledge of the practice of preconceptive sex selection will do much extra psychic damage. (Late sex-selective abortion may be a different story, since the killing of sentient female fetuses because of their sex is a far more drastic expression of sexism.)

Women and Patriarchal Medicine

The new reproductive technologies are being developed and marketed primarily by men and for the profit of men. Yet it is primarily women and children who undergo the medical risks and suffer any harmful side effects. These technologies are part of the historical process by which men have gradually extended their control of human reproduction. The various home recipes for sex selection are probably almost totally ineffective. The only reliable method currently in existence, i.e., prenatal sex diagnosis and selective abortion, requires the woman to submit to repeated medical interventions and involves significant risks. Sex selection by insemination with sex-separated sperm also requires the assistance of professionals, most of whom are men. And if a sex-selection pill or injection is developed it is a safe bet that it will be administered to women while the proceeds go to men.

We are coming to understand just how much women have paid for the male monopoly of the medical profession. Even the most empathic of men is ill-equipped to provide gynecological or obstetric care which accords with women's own felt needs; for he has never had any of the experiences associated with having a female reproductive system, or with being a woman in a patriarchal society. Too often, he is ready to undertake dangerous surgical or pharmacological interventions with an alacrity which would be unthinkable in the case of a male patient. Unnecessary hysterectomies, mastectomies, and Caesarian sections, involuntary sterilizations, carcinogenic estogen replacement therapy, and the thalidomide, DES, and Dalkon Shield disasters are just a few of the results of this attitude.

The male monopoly of female health care is a relatively recent phenomenon. Before the eighteenth century, midwifery was almost entirely a female profession. Barbara Ehrenreich and Deirdre English's classic study, *Witches, Midwives and Nurses: A History of Women Healers*, describes the gradual suppression of female medical practitioners, from the mass murders of the witch-craze era—many of the victims of which were female healers—to the eventually successful drive to outlaw the practice of midwifery other than by members of the male medical profession.[24] Prior to the end of the nineteenth century, women in most of the western world were legally prohibited from studying or practicing medicine. The methods of male physicians were no safer or more effective than those of female healers— quite the contrary. Mary Daly notes "the stunning reversal which gynecological historians have inflicted upon our minds by referring to the "filthy" midwives who were replaced by antiseptic ob/gyns."[25] In reality, male physicians caused more deadly contagion than female midwives had ever done.

> In the seventeenth century began a plague of puerperal fever which was directly related to the increase in obstetric practice by men ... the hands of the physician or of the surgeon, unlike those of the midwife, often came directly from cases of disease to cases of childbirth ... With the growth of lying-in hospitals in the cities of Europe the disease—rarely known in earlier times—reached epidemic proportions ... in the month of February 1866 a quarter of the women who gave birth in the Maternité hospital in Paris died.[26]

Eventually, the male medical profession discovered the need for asepsis, and the plague of "childbed fever" came to an end—to be replaced by "the age of anesthetized, technologized child-birth."[27] But thousands of women continued to die from illegal abortions, and many more were forced to bear children against their will because of antiabortion laws. In the United States, these laws were largely a result of the campaign conducted by the organized medical profession.[28] Moreover, through its opposition to what it regards as socialized medicine, the medical profession has repeatedly blocked reforms which could have improved the delivery of health-care services to poor women and their children.[29]

Thus, it is not surprising that the new reproductive technologies put many women in mind of the saying about Greeks bearing gifts. In the past, "men ... gradually annexed the role of birth-attendant and thus assumed authority over the very sphere which had originally been one source of female power and charisma."[30] Now they are assuming additional control of the biological processes of conception and gestation. Unless women become more involved in the development and administration of these new technologies, sex selection, in vitro fertilization, and the like may "only add to the persistent medicalization of women's bodies, and thus to the biomedical management of women's lives."[31]

So long as most women must rely upon male physicians for the means of controlling their own reproductive lives, we cannot be confident that the new reproductive technologies will be developed and made available in ways that respect women's needs. Emily Erwin Culpepper speaks of "the enormous ego involvement in these technologies that the men developing and using them have."[32] She notes that,

> These technologies function as male extensions of themselves, an observation that leads [some] feminists to consider them as sophisticated forms of rape, battering and mutilation of women.[33]

But none of these observations constitutes an objection to sex selection as such. There are feminists who share these critical views of the male-dominated medical profession, yet who advocate the use of sex selection to produce daughters. If more women decide that sex selection is consistent with their feminist principles, there may soon be feminist-run sperm banks offering sex-selected sperm. There is already at least one feminist-run sperm bank, operating in Oakland, for the benefit of women who wish to have children without a male partner. Women's health-care collectives may hire physicians to provide sex-selection services under conditions specified by the women themselves. Such strategies may provide women with freer and less compromising means of access to the more effective means of sex selection.

In the long run, however, women will achieve control of their own reproductive lives only when the male monopoly of the practice of medicine and medical research is ended. The painfully slow increase in the numbers of female physicians has thus far not been enough to alter fundamentally the attitudes and practices which subvert the autonomy of the female patient. There are still institutional barriers to the entry of women into the practice of medicine. One such barrier is the typical requirement that medical school, internship, and residency be undertaken on a full-time basis, requiring extremely long hours which are impossible for most women who have children. The medical profesion must be altered to facilitate the entry of women of all ages and ethnic groups, and to enable the recipients of health-care services to function as active consumers rather than passive patients. Women from all socioeconomic, age, and ethnic groups need to be proportionately represented in all medical school faculties, hospital administrations, and government agencies which allocate money for research and health-care delivery. Where it has not already been done, health care must be "socialized"—i.e., made available to all who need it, regardless of their ability to pay.

Just as fundamentally, the health-care monopoly of the standard medical profession needs to be broken. And it is being broken, in a number of ways. Medical self-education and self-care, female-operated fertility control clinics, the licensing of midwives and other paramedical profesionals, and the spread of acupuncture, herbal medicine, homeopathy, and other alternative forms of treatment are all part of this process. Alternative systems of health care will have to prove their safety and efficacy—but so, increasingly, will the methods offered by the mainstream practitioners. Only as such cumulatively revolutionary changes occur will women be able to welcome the new reproductive technologies without the fear that they will reduce our reproductive freedom rather than extending it.

The Dilemma of Womanless Men

Thus far we have focused upon the possible ill effects of sex selection upon women. But men might also be adversely affected were sex ratios to increase significantly. Many heterosexual men may be unable to find female partners at all. For a time, this problem might be alleviated by marriages between younger men and older women, who at present are often unable to find male partners. But if males become a majority at all age levels, this solution will no longer suffice.

On the basis of surveys of American sexual preferences done in the 1940s and 1950s, Amitai Etzioni estimates that the general practice of sex selection would result in about a 7 percent increase in the relative number of males born. This, added to the existing biological pattern of about 51 percent male births, would produce a sex ratio at birth of about 109. Hence, there might be something on the order of 360,000 "surplus" males born in the United States each year. Even if this figure proves inflated, it is likely that in time there will be millions of men "who will not find mates and who will have to avail themselves of prostitution, homosexuality, or be condemned to enforced bachelorhood."[34] Where son-preference is stronger than in America, the surplus of males might be even greater.

Whether this is regarded as a good or a bad result for men will obviously depend upon whether one regards monogamous heterosexual marriage as better for men than the alternatives. Etzioni clearly does. In his view, the sorrows of unmated men would greatly outweigh the joys of parents who would be enabled to have children of the sex they prefer. To some, this assumption may seem implausible. Bachelorhood is often depicted as a paradisiacal condition compared to that of the married man. It is usually women who are assumed to need marriage.

But, as Jessie Bernard points out, marriage has not deserved the bad press which it has had among men. Summarizing the relevant research, she notes that married men are superior to never-married men on almost every social, psychological, or demographic index.[35] The average married man, compared to the average bachelor, has greater earning power, better mental and physical health, and a longer life span. The suicide rate for single men is almost twice as high as for married men. Married men commit fewer crimes than single men, and are far less apt to be institutionalized for psychiatric problems. Most men do not reject marriage, once they have experienced its benefits:

> Most divorced and widowed men remarry. At every age, the marriage rate for both divorced and widowed men is higher than the rate for single men. Half of all divorced white men who remarry do so within three years after divorce. Indeed, it might not be far-fetched to conclude that the verbal assaults upon marriage indulged in by men are a kind of compensatory reaction to their dependence on it.[36]

Bernard is aware that not all of the credit for the happier circumstances of married men goes to marriage itself. Women tend to marry

men who are somewhat "above" them in age, education and occupation; consequently, never-married men tend to be "bottom-of-the-
barrel" in terms of various socioeconomic measures.[37] Yet comparisons between married and never-married men of similar occupational
status also show better health and longevity among the married.[38]
Men, it seems, do benefit from marriage; and it is no mystery why
they do. It is conducive to a man's health and happiness to have "a
wife well socialized to ... devote her life to taking care of him,
providing ... the regularity and security of a well-ordered home."[39]
Indeed, marriage may be becoming more beneficial for men. As more
women contribute to the family income, while continuing to do most
of the housework and child care, husbands are gaining freedom from
exclusive economic responsibility, without taking on much extra
domestic work.

Unfortunately, marriage is not so good a bargain for women. Single
women suffer less than either single men or married women from
anxiety, depression, and feelings of inadequacy. Married women who
have paid jobs earn less on the average than single women, and suffer
more from mental and physical health problems. Housewives suffer
most of all from psychological and psychosomatic ailments, which are
an index of their discontent.[40] The inferior health and happiness of
the average married woman, like that of the average unmarried man,
is not entirely due to the effects of marriage itself. Independent,
self-assured women are probably less apt to be attracted by marriage,
and are less attractive to many men. Through this mutual selection
process, many of the strongest and most occupationally successful
women remain single.

Yet the problems of married women are in large part due to the
conditions of patriarchal marriage. In becoming housewives, or in
accepting the double burden of housework and paid labor, women
often sacrifice education, social contacts, occupational prospects,
personal independence, health, and ultimately self-respect. In many
jurisdictions, marriage still involves a loss of legal rights for women.
Worse, it often entails a loss of identity. The wife not only loses her
name, but is expected to reshape her personality and her life to
conform to her husband's needs or demands.

To the extent that men benefit from marriage at women's expense,
it is tempting to respond to the objection that higher sex ratios will
deprive many men of these benefits by saying, "Tough luck!" Men

have both the power and the moral obligation to so alter the terms of marriage and other institutions that the reasons for son-preference will be swept away. If they do not, then they have no grounds for complaint if the implementation of son-preference deprives some men of the chance to have wives of their own.

But this would be an unsympathetic response. Men are not all individually resonsible for the injustices of patriarchy, and many of the unhappy bachelors whom Etzioni imagines would probably be innocent non-chauvinists. But would the wifeless state really be such an unfortunate one for the "excess" men of the future? It is not obvious that monogamous heterosexual marriage will continue to be the most popular form of life—still less that it ought to. There have always been some men and women who have found homosexuality, celibacy, or nonmonogamy a superior alternative. We do not know that any particular pattern of sexual behavior is the most natural one for most human beings. All sexual behaviors (except perhaps mastur-bation) must be socially learned by members of our species.

It is possible that, in the absense of social pressures towards particular forms of sexuality, most people would remain what Freud termed "polymorphous perverse"—i.e., open to many forms of erotic pleasure, not all of them centered on genital contact. The assumption that most people are naturally heterosexual is unproven and unpro-vable so long as there are other likely causes of the prevailing heterosexual orientation present. And there *are* other likely causes present. Heterosexuality is compulsory for both women and men in most contemporary patriarchal societies. Homosexuality is punished in a myriad of ways, from ridicule or the loss of employment to imprisonment, torture, or death.

Compulsory heterosexuality has long served as a butress for other patriarchal institutions. By establishing "the" sex act as a means by which men prove their manhood through the conquest of women, it plants the model of male domination and female submission in the fertile soil of human eroticism. By denying the erotic love of women for women and of men for men, it makes it more difficult for women to unite in opposition to their common oppression, or for men to bond with other men, except in warlike enterprises directed towards the conquest of some enemy. And by focusing eroticism upon heterosex-ual genital intercourse, it binds women and men to marriage and child rearing.[41] All this may once have been survival-conducive—

particularly for patriarchal societies in competition with other patriarchal societies. Where warfare is inescapable, an expanding population may be a vital asset. Perhaps this is why so many patriarchal societies have sought to suppress nonprocreative forms of sexual behavior.

But continued population growth is no longer advantageous. As the need to limit the number of births becomes more apparent, the rationale for compulsory heterosexuality grows thinner. In many nations, gay liberation movements have already removed some of the stigma from the male homosexual and lesbian forms of life—though the current scare over AIDS, which is often spread by male homosexual intercourse, threatens to reverse that progress. They have demanded not only that laws criminalizing consensual homosexual activities between adults be repealed, but that marriage between same-sex persons be granted the same legal and social recognition as heterosexual marriage. These demands are both just and necessary. The rationale for the prohibition of polyandry, polygyny, and other forms of nonmonogamous marriage has also worn thin, and will also be challenged.

It is also likely that same-sex couples or groups, and single persons, will secure both the legal right and the technical means to have and rear children of their own, without procreating in the natural way. Artifical gestation, parthenogenesis, ovular merging, and cloning are possible developments which could help to undermine the heterosexual monopoly of parenthood. As heterosexuality becomes a matter of choice, rather than a socially inculcated compulsion or a necessary condition for legitimate parenthood, numerical imbalances between women and men will no longer inevitably mean that some portion of the population will be unable to find lovers, or to enjoy the pleasures of family life.

Of course, we do not know that any such fundamental changes in the social shaping of human sexuality will occur. Perhaps the hegemony of heterosexuality will never be much further eroded than it has been already. Perhaps Etzioni's prophecy of millions of unhappy bachelors will come true, if sex selection comes into widespread use. But the emergence of viable alternatives to heterosexual marriage is more than a fantasy. It will not come as a spontaneous result of the new reproductive technologies, but will require fundamental changes in deeply entrenched attitudes and repressive laws.

Legislators and suppliers of reproductive technologies will continue to be subject to pressure from conservatives to make these technologies available only to married heterosexual couples. Feminists and civil libertarians will have to exert an even stronger pressure in defense of the right of individuals to conduct their reproductive lives in accordance with their own moral convictions—so long as they do not violate the rights of other persons. The more we succeed in this, the less we will need to worry that sex selection will produce cohorts of miserable mateless men.

Increased Prostitution

Another fear is that if women become scarce due to the use of sex selection and many men are unable to find wives, there will be more female and male prostitutes to service the male demand for sexual outlets. This is a real possibility, but it is not a strong objection to sex selection, for several reasons.

First, marriage is no guarantee that men will not seek the services of prostitutes. Many prostitutes report that a majority of their customers are married men. Second, there is no automatic causal relationship between high sex ratios and high rates of prostitution. Where prostitution is genuinely contrary to the prevailing ethos— and not just hypocritically said to be so—it may be uncommon even when sex ratios are high. Conversely, where sex ratios are low but many women and young men are impoverished, both male and female prostitution may flourish.

Third, the sale of sexual services is not necessarily a bad thing, either for the vendor or the customer. Many feminists believe that female prostitution degrades all women, by perpetuating the assumption that women's role is to service male sexual desires.[42] There is some truth to that theory. But many of the evils associated with prostitution are more a function of its illegality than of the activity itself.[43] Where the sale of sexual services is a legal profession, it need not be degrading. Many female prostitutes argue that their profession is less degrading than patriarchal marriage. Prostitutes may have more economic independence than most wives, and more opportunity to choose when and to whom they will provide sexual services. Where prostitution is legal and prostitutes have access to sanitary facilities and medical services, it need not greatly increase the spread of venereal diseases—although this is certainly a risk, in the age of AIDS and antibiotic-resistant bacteria.

But the most telling response to this objection is that women (and men) usually resort to prostitution only when they have few equally satisfactory means of earning a livelihood. The best way to minimize prostitution (assuming that this is a desirable goal) is not to try to prevent sex-ratio increases, but to provide education and employment opportunities which will be more attractive to women and young men than the sale of sexual services.

Masculinism and the Rape of the Natural World

There is a deep symbolic association between the subjugation of women under patriarchy and the progressive destruction of the natural world. In the western tradition, nature has been conceptualized as feminine, while women have been regarded as a part of nature—in contrast to "mankind," which is assumed to be outside of and above nature.[44] In this masculinist world view, women and nature exist to be exploited by men. Although women may profit from the destructive exploitation of the natural world, most direct acts of destruction—from the extinction of animal species to the pollution of the land, air and water with industrial contaminants—have been the work of men.

Elizabeth Dodson Gray has argued that it is the male lack of certain female experiences which leads men to wantonly destroy the nonhuman elements of nature. In her view men, because they neither menstruate nor bear children, tend to have a weaker sense of their connection with the rest of nature and with the future of our own species. Thus, she says,

> No matter how androgynous men may become, it is ... not possible for men alone to lead us into a society with a fully developed sense of its limited but harmonious place in nature. It is not possible for men alone to do this because the male's is simply a much diminished experience of body, of natural processes, and of future generations.[45]

If this analysis is correct, sex-ratio increases may bode ill for the future of nonhuman life forms on this planet. But it is doubtful that the destructive behavior of modern man towards the rest of nature is due as much to male biology as to certain elements in the western cultural tradition. This tradition—though developed primarily by and for males—is in no way necessitated by the male's lack of the experiences associated with having a female reproductive system. J. Baird Callicott contrasts the standard western conception of nature as innert and soulless with that of many aboriginal American peoples, who have traditionally regarded all the elements of nature as alive and possessed of spirit. In this world view, animals, plants, and other nonhuman entities are entitled to a respect akin to that which is due to human persons. Rather than being regarded as passive matter, available for human exploitation, they are "conceived to be coequal members of a natural social order."[46]

> The American Indian lived not only by a tribal ethic, but by a land ethic as well, the *overall* and *usual* effect of which was to establish a greater harmony between Indians and their environment than enjoyed by their European successors.[47]

It may be that such a land ethic could not have arisen in a society as patriarchal as our own, in which women's experiences have little place in the dominant cultural perspective. Nevertheless, the fact that both men and women of other societies have lived by such a nonanthropocentric ethic shows that male psychobiology cannot account for the environmental destructiveness of contemporary industrialized nations. Male biology alone does not prevent the growth of environmental awareness. The lack of female experiences has not prevented men like Aldo Leopold from understanding that humanity must (re)learn how to exist as a part of the natural community.[48] Furthermore, high sex ratios do not necessarily produce environmentally destructive cultures. There have been high-sex-ratio societies (such as those of the Eskimos) which have existed in balance with the natural world for thousands of years.

The feminist and ecological movements are philosophically and spiritually linked. The same masculinist world view underlies both the oppression of women and the destruction of the nonhuman world. Thus, neither of these movements can succeed in isolation from the other. The backlash against feminism strikes at the environmentalist movement as well. But not all men accept this masculinist world view, and not all women reject it. There are both women and men on both sides of the debate over "development" versus the preservation of remaining natural species and habitats. Thus, it is impossible to know that, if sex ratios increase due to sex selection, the irresponsible destruction of the natural world will be more likely to continue.

Will Class Differences Be Accentuated?

Another possible ill effect of sex selection is the aggravation of economic class differences and class conflict. Sex-ratio changes due to the use of sex selection are likely to differ between social classes and ethnic and religious groups, for both economic and ideological reasons.

In some nations, son-preference is negatively correlated with socioeconomic class. In a 1956 study of Italian couples, Giurovich reported that son-preference is stronger among lower-class couples. He surmised that this is because sons are more conducive to the family's upward economic mobility.[49] In an American study of 1965, Lee Rainwater found that white upper-lower class couples were most apt to spontaneously express a preference for sons, while white upper-middle class couples were least apt to do so.[50] Gerald Markle, on the basis of a survey conducted in the Tallahassee, Florida area, concluded that son-preference tends to decline with increased education. He found that,

> Respondents with no more than a high school education would prefer almost five males for every non-male, compared with male ratios of 2.2 and 1.1, respectively, for respondents with college education and with advanced college degrees.[51]

Such findings have led some to argue that if sex selection were to be made widely available, sex-ratio increases would be greater among lower socioeconomic status groups, and that existing social tensions would therefore be aggravated. Etzioni says that,

> Interracial and interclass tensions are likely to be intensified because some groups, lower classes and minorities specifically ... seem to be more male oriented than the rest of the society ... This may produce an especially high boy surplus in lower status groups. These extra boys would seek girls in higher status groups (or in some other religious group than their own)—in which they also will be scarce.[52]

As an objection to the use of sex selection, this argument has two major weaknesses. In the first place, it is not obvious that the effects of sex-ratio increases among the poor would be entirely negative. Poor women might benefit from an increase in dyadic power, and there might be fewer poor women forced to raise their children without male support. A relative increase in the number of men born poor would not automatically increase the economic power of men, as would an increase in the number of males in the wealthy classes.

In the second place, the prediction of higher sex-ratio increases among the poor as a result of sex selection is questionable. This prediction depends upon the assumption that when effective means of sex selection become available, they will be equally accessible to all social classes. This may be the case, if the new methods are inexpensive and easy to distribute on a large scale, or if their use is subsidized through socialized medical programs. But sex selection may prove to be expensive, and may require sophisticated technologies which cannot be made generally accessible. Furthermore governments, fearing the effects of sex-ratio increases, may refuse to subsidize its use, or even seek to ban it. In such circumstances, sex selection will remain a prerogative of the relatively wealthy, and thus it will probably be the upper classes which experience the greatest increase in sex ratios.

Disproportionate sex-ratio increases among the more privileged classes would certainly not be conducive to sexual equality. As Roberta Steinbacher points out, this would mean "that increasing numbers of women in the future are locked into poverty while men continue to grow in numbers in positions of control and influence."[53] Increased wealth and power in the hands of men could only result in the aggravation of the entire range of injustices against women. More women might be able to marry upward, but the resulting socioeconomic inequality between many married couples might intensify the perception of female inferiority.

Yet we cannot move directly from these facts to the conclusion that sex selection is immoral. What is immoral is that it should be made available only to the wealthy. If we want to avoid some of the worst potential social consequences of sex selection, we must make it equally available to all social classes, while at the same time continuing to oppose the conditions which create son-preference.

Are We on a Slippery Slope?

Some people fear that sex selection and other new reproductive technologies are the initial steps down a road that will lead to moral atrocities. Some fear, for instance, that totalitarian governments will mass-produce human beings genetically engineered to suit their nefarious purposes. Others worry less about the abuse of reproductive technologies by governments than about the effects of their voluntary use by individual parents. Helen Holmes argues that,

> Selection of children's sex starts us on a slippery slope. What traits would we like to be able to specify in our children? Hair color? IQ? The physique for our favorite sport? For many parents such traits in their children are more important than sex... If we are going to custom-design our children, for which traits is there moral justification?[54]

All slippery slope arguments presuppose that people cannot make certain distinctions which the arguer considers vital. For if the relevant distinctions can be made, then there is no reason to suppose that the acceptance of the one behavior will lead to the acceptance of the other. Such arguments fail if either (1) people can make such distinctions, or (2) these distinctions do not have the significance which the arguer takes them to have. In this case, both objections apply. Many people who do not object to sex selection would object to the preselection of children's hair color or IQ, because they perceive that these cases involve quite different considerations. Most of the arguments for and against the preselection of sex would not apply to the preselection of hair color, which usually has much less social significance. Preselecting for intelligence would raise much more serious moral questions, because intellectual ability (like sex) often has a very significant effect upon a person's life prospects. These cases can and must be treated separately.

The fear that sex selection will lead to more objectionable practices in the future is sometimes based on a fear of *all* forms of so-called positive eugenics. Negative eugenics, i.e., the attempt to prevent the birth of children with severe illnesses or abnormalities, is generally accepted as morally appropriate and even morally obligatory. But positive eugenics, i.e., selection in favor of some nonpathological traits in preference to others, may remind us of the genocidal crimes of the German Nazi regime. Holmes maintains that,

> Positive eugenics, that is, the deliberate selection of genetic traits for human beings coming into this world, is morally wrong. After all, we humans are not really wise enough to know what traits will be best for the good of humankind through all eternity ... Moreover, the Nazi eugenics program clearly demonstrated the diabolical nature of eugenics policies.[55]

But should we condemn *all* forms of positive eugenics on the grounds that *some* forms of positive eugenics are morally atrocious? Some forms of positive eugenics might provide clear and significant benefits to future persons. Suppose, for instance, that there were a safe and inexpensive preconceptive or prenatal procedure which would so alter a child's genetic constitution as to guarantee a lifetime of excellent vision, or an immunity to all viral or bacterial infections. (This would be positive, not negative eugenics, since most normal people today do not have excellent vision throughout their lives and are not immune to all infections.) Arguments might arise about whether such genetic "improvements" would really be desirable in the long run. But I see no a priori reason to deny that they might be.

What of the argument that eugenic technologies might be abused by totalitarian governments? Indeed they might. If we fear that genetic engineering will eventually lead us to the Brave New World scenario, then perhaps we ought to draw the line at selecting sex (which does not involve tampering with the genes), rejecting all more radical forms of positive eugenics. But I believe it would be wiser to concentrate on the defense of political freedoms and human rights; for it is the lack of these which creates the danger that new biomedical technologies will be abused by totalitarian governments. As Peter Singer and Deane Wells point out,

> The fact that we have not yet mastered ... genetic engineering is ... not the only thing that distinguishes our society from Huxley's disutopia ...
> The most odious aspects of Huxley's world are undoubtedly its caste system and its lack of freedom. These have existed in many societies, and indeed still do exist in some, without any recourse to biotechnology. The human race does not need advanced biotechnology to aid it in developing odious social systems.[56]

Positive eugenics may be feared for a number of reasons: it might be used for immoral purposes; it might prove to have unforeseen side effects; it might divert medical resources from more important purposes; and, above all, the advocacy of eugenics has been historically associated with vicious racist doctrines. There are sound reasons for proceeding cautiously, with full public disclosure and extensive public discussion of each new or proposed procedure—just as ought to be done in every other area of medical research. They are not, however, reasons for a rejection of all forms of positive eugenics, including sex selection (if that is viewed as a form of positive eugenics). Each new or proposed technique will have to be evaluated individually, through extensive public discussion of its costs, risks, and potential benefits. Many will prove too dangerous, too expensive, or too morally questionable to be worth pursuing. Perhaps some of the new means of selecting sex will be among the options which will have to be rejected. But if we refuse to make the necessary distinctions, insisting that all forms of positive eugenics must be accepted or rejected as part of a single package, we may inadvertently contribute to the very sorts of abuses which we fear.

Notes

1. Marcia Guttentag and Paul F. Secord, *Too Many Women? The Sex Ratio Question* (Beverly Hills: Sage Publications, 1983).
2. Guttentag and Secord, 23.
3. Guttentag and Secord, 37–52.
4. Guttentag and Secord, 114.
5. Guttentag and Secord, 29.
6. Anne Summers, *Damned Whores and God's Police: The Colonization of Women in Australia* (Blackburn, Victoria: Penguin Books, 1975), 267–71.
7. Guttentag and Secord, 28–29.
8. See Andrea Dworkin, *Right Wing Women* (London: The Woman's Press, 1983).
9. See, for instance, Robyn Rowland, "Reproductive Technologies: The Final Solution to the Woman Question?" in *Test Tube Women: What Future for Motherhood?*, edited by Rita Arditti, Renate Duelli Klein, and Shelly Minden (London: Pandora Press, 1984), 361.
10. Sir Francis Galton, *English Men of Science: Their Nature and Nurture* (London: Frank Cass and Company, 1970).
11. See, for instance, W.D. Altus, "Birth Order and Its Sequelae," *Science* 151 (January 1966), 44–49.
12. Bert N. Adams, "Birth Order; A Critical Review," *Sociometry* 35:3 (1972), 411.
13. Alfred Adler, *Understanding Human Nature* (New York: Greenberg Publisher, 1927); "Characteristics of the First, Second and Third Child," *Children* 3:14 (1928); and *What Life Should Mean to You* (Boston: Little, Brown, 1931).
14. Robert B. Zajonc, "Dumber by the Dozen," *Psychology Today*, January 1975, 37–42; and R. B. Zajonc, Hazel Markus, and Gregory B. Markus, "The Birth Order Puzzle," *Journal of Personality and Social Psychology* 37:8 (1979), 1325–41.
15. Zajonc (1975), 42.
16. Lucille K. Forer, *The Birth Order Factor: How Your Personality Is Influenced by Your Place in the Family* (New York: David McKay Company, 1976), 104.
17. Bradford Wilson and George Eddington, *First Child, Second Child: What Your Birth Order Means to You* (London: Souvenir Press, 1982), 85.
18. Cécile Ernst and Jules Angst, *Birth Order: Its Influence on Personality* (New York: Springer-Verlag, 1983), 45.
19. Carmi Schooler, "Birth Order Effects: Not Here, Not Now!" *Psychological Bulletin* 78:3 (September 1972), 161–75.
20. Ernst and Angst, 13.
21. Ernst and Angst, 187.
22. Roberta Steinbacher, "Futuristic Implications of Sex Preselection," *The Custom-Made Child? Women-Centered Perspectives*, edited by Helen B. Holmes, Betty B. Hoskins, and Michael Gross (Clifton, New Jersey: Humana Press, 1981), 187.
23. John C. Fletcher, "Is Sex Selection Ethical?" *Research Ethics* 128 (1983), 343.
24. Barbara Ehrenreich and Dierdre English, *Witches, Midwives and Nurses: A History of Women Healers* (Old Westbury, New York: The Feminist Press, 1973).
25. Mary Daly, *Gyn/Ecology: The Metaethics of Radical Feminism* (Boston: Beacon Press, 1978), 236.

26. Adrienne Rich, *Of Women Born: Motherhood as Experience and Institution* (London: Virago, 1977), 151.

27. Rich, 155.

28. See James C. Mohr, *Abortion in America: The Origins and Evolution of National Policy, 1800-1900* (New York: Oxford University Press, 1978).

29. For instance, in 1929 the organized opposition of the American Medical Association led to the repeal of the Sheppard Townes Act, which for a decade had provided support for maternal and infant health care. See Beverly Wildung Harrison, *Our Right to Choose: Toward a New Ethic of Abortion* (Boston: Beacon Press, 1983), 172.

30. Rich, 129.

31. Janice Raymond, "Sex Preselection," *The Custom-Made Child?*, 212.

32. Emily Erwin Culpepper, "Sex Control, Science and Society," *Science* 161 (September 1968), 307.

33. Ibid.

34. Amitai Etzioni, "Sex Control, Science and Society," *Science* 161 (September 1968), 1110.

35. Jessie Bernard, *The Future of Marriage* (London: Souvenir Press, 1973), 17.

36. Bernard, 18.

37. Bernard, 33.

38. Bernard, 19.

39. Bernard, 24.

40. Bernard, 30.

41. See Adrienne Rich, "Compulsory Heterosexuality and Lesbian Existence," *Signs: Journal of Women in Culture and Society*, 5:4 (Summer 1980), 631-60.

42. See, for instance, Susan Brownmiller, *Against Our Will: Men, Women and Rape* (New York: Simon and Schuster, 1975), 390.

43. Also see Chapter 8, third section.

44. See, for instance, Susan Griffin, *Woman and Nature: The Roaring Inside Her* (New York: Harper and Row, 1978).

45. Elizabeth Dodson Gray, *Green Paradise Lost* (Wellesley, Massachusetts: Roundtable Press, 1979), 114.

46. J. Baird Callicott, "Traditional American Indian and Traditional Western European Attitudes Towards Nature: An Overview," *Environmental Philosophy: A Collection of Readings*, edited by Robert Elliot and Arron Gare (St. Lucia: University of Queensland Press, 1983), 249.

47. Callicott, 250.

48. Aldo Leopold, *A Sand County Almanac* (New York: Oxford University Press, 1949), 204.

49. G. Giurovich, "Sul Desiderio dei Coniungi di Avere Figle e di Avere Figle di un Dato Sesso" ("On the Wish of Married Couples to Have Children and to Have Children of a Specified Sex"), *Atti Della 16 Riunione Scientifica della Societa Italiana di Statistica* (1956), Rome, 287-317.

50. Lee Rainwater, *Family Design* (Chicago: Aldine, 1965), 131.

51. Gerald E. Markle, "Sex Ratio at Birth: Values, Variance, and Some Determinants," *Demography* II:1(February 1974), 135.

52. Etzioni, 1109.

53. Steinbacher, 188.

54. Helen B. Holmes, "Sex Preselection: Eugenics for Everyone?" forthcoming in *Biomedical Ethics Reviews, 1985*, edited by James Humber and Robert Almeder (Clifton, New Jersey: Humana Press, 1985).

55. Ibid.

56. Peter Singer and Deane Wells, *The Reproduction Revolution: New Ways of Making Babies* (New York: Oxford University Press, 1984), 43.

The Possible Benefits

The consequentialist arguments against sex selection are debatable, but sufficient to show that it *might* have some undesirable consequences, particularly for women. On the other hand, in some circumstances it might also have some beneficial effects, such as increasing women's dyadic power in heterosexual relationships. Selective abortion is already used to prevent the birth of children with severe sex-linked diseases. Those who have argued that more effective methods of sex selection ought to be developed and made available have pointed to several other possible benefits. These include: (1) both short-term and long-term reductions in the birth rate; (2) greater happiness for parents; and (3) avoidance of the birth of children who are unwanted because of their sex. I would add a fourth possible benefit: sex selection will provide some women with new means of resisting patriarchy.

Some of these benefits are just as uncertain as are the feared detrimental effects of sex selection. Nevertheless, they must be taken into account if we are to have a balanced view of its possible social consequences.

Preventing Sex-Linked Diseases

The only widely accepted use of sex-selective abortion in the industrialized nations is to prevent the birth of male children with severe sex-linked diseases. Such diseases are caused by defective genes locted on the X-chromosome, and thus are sometimes called "X-linked" diseases. Because the trait is recessive, it is rarely manifested in female carriers of the defective gene, who usually have a normal gene on the other X-chromosome. Male carriers will pass the defective gene to daughters but not to sons, to whom they do not contribute an X-chromosome. Female carriers will pass the defective gene to both daughters and sons but, unless the father also carries the defective gene, the disease will be manifested only in sons. Thus, about half of the sons of female carriers, but almost none of the children of male carriers, will have the disease.

Some X-linked abnormalities are not serious, and would rarely prompt anyone to consider abortion. But others, such as hemophilia and Duchenne's muscular dystrophy, are severe and chronic diseases which are very often lethal. Because these X-linked diseases cannot be detected prenatally, at present the only way that women who may be carriers can avoid the risk of having sons who will have the disease is by either having no children at all, or using amniocentesis and the selective abortion of male fetuses to produce an all-girl family. Effective methods of preconceptive sex selection, or of detecting fetal sex earlier in the pregnancy, would eliminate the need for second-trimester abortions following amniocentesis. Preconceptive sex selection would also enable women who carry severe X-linked diseases but who are unwilling to undergo abortion to have children without the risk of having affected sons.

The incidence of the more severe X-linked diseases is not high, and the total number of late abortions performed to avoid such diseases is probably not more than a few hundred per year in the United States. However, the importance of a method of avoiding X-linked diseases without late abortion should not be minimized. Such abortions can be extremely traumatic for the woman and for those close to her. Rayna Rapp has described her own experience of the abortion of a second-trimester fetus diagnosed as having Down's syndrome. (Down's syndrome is caused by a chromosome abnormality which does not appear to be hereditary. It produces mental retardation, and a range of physical abnormalities which often result in early death.) Rapp speaks of the psychological pain of having to end the life of a fetus which she had wanted and had already carried for almost five months, and of the physical trauma of undergoing late abortion and the recovery from it.[1] She did not regret her decision, and neither do most women carriers of X-linked diseases who choose to abort male fetuses regret that choice. As Laurence Karp points out,

> The large majority of carrier women have closely followed the progress of the disease in a brother or cousin, and they are certain that they do not

want to take a 50 percent, a 25 percent, or even a 12½ percent chance that their own sons will suffer from the disease.[2]

Nevertheless, these women and their families would benefit greatly from preconceptive sex selection or earlier fetal sex diagnosis, which would avoid the need for late abortion.

Besides the trauma of late abortion, there are other problems with the use of amniocentesis and selective abortion to avoid X-linked diseases. First, about half of the male fetuses which are aborted are normal. Second, about half of the female fetuses which are brought to term will carry the defective gene. When these girls grow up they will probably face the same problem of whether or not to have children and risk passing the defective gene to those children.[3] Thus, it can be argued that it would be better for people who know that they (may) carry the gene for a severe X-linked disease to voluntarily remain childless. It is not clear, however, that they are morally obligated to avoid having children, if those children will not themselves be afflicted with the disease.

Some ethicists object to the selective abortion of abnormal fetuses on the grounds that it is apt to make people less sympathetic towards those who suffer from similar abnormalities. Leon Kass, for instance, asks,

> How will we come to view and act toward the many "abnormals" that will remain among us—the retarded, the crippled, the senile, the deformed, and the true mutants—once we embark on a program to root out genetic abnormality? ... the idea of "the unwanted because abnormal child" may become a self-fulfilling prophecy, whose consequences may be worse than those of the abnormality itself.[4]

This objection is implausible for the same reason that the "slippery slope" objection to other new reproductive technologies is implausible. There is no reason to believe that those who seek to avoid the birth of children with severe genetic diseases will become unsympathetic to the needs of people who suffer from such diseases. Indeed, it is largely sympathy for such people which motivates efforts to avoid the birth of children who will suffer in the same ways.

There are, however, grounds for concern about the psychological effects of "eugenic" abortions—especially late ones—upon people who have the same ailment as the fetuses which are aborted. Just as women may suffer from the knowledge that sentient female fetuses are being killed because of their sex, so victims of serious diseases such as hemophilia may find it painful to know that sentient fetuses are being aborted because they may have the same disease. So long as effective methods of preconceptive sex selection are not available, and so long as not all persons who carry the genes for serious X-linked diseases are willing to remain childless, such abortions will sometimes be justified. But the earlier they can be done the better it will be for all concerned; and it would surely be better if sex could be selected before conception.

On the other hand, it is possible that the use of either preconceptive sex selection or sex-selective abortion for the prevention of most sex-linked diseases will soon become unnecessary. As Holmes points out,

> Recombinant-DNA technologies are snowballing. Soon there may be created radioactive DNA probes for all the common and many of the rare genetic diseases. Such a probe would bind to the defective gene in the DNA isolated from fetal cells obtained by amniocentesis or chorionic biopsy. When the *gene itself* can be detected, only a son who *actually has* the genetic disease need be aborted.[5]

In the meantime, however, the prevention of sex-linked diseases, without late abortion, would be a significant benefit of new methods of selecting sex. The next alleged benefit is much wider in its scope, but more debatable in its impact.

Birth-Rate Reductions

The human population of our planet has grown enormously in the past several centuries. It is estimated that one million years ago there were about 125,000 human beings, and that the rate of population growth was about 2 percent per millennium. The development of agriculture produced the first rapid acceleration of population growth. However, it took about 10,000 years for the population to reach one billion, early in the nineteenth century. The scientific-industrial revolution enabled the second billion mark to be reached by 1930. Since then the population has grown to about 4.75 billion. This extraordinary growth is not due to higher birth rates, but to such factors as increased food production and the near elimination of many lethal diseases which formerly kept death rates closer to birth rates.

By a sad irony, population growth tends to be highest in those nations which can least affort it. The industrialized nations have been undergoing a transition to lower birth rates for several centuries. Many are approaching, or have already reached, the level of zero population growth—at least if we discount the effects of immigration. But in many of the impoverished nations of Africa, Asia, and Latin America, growth continues at a rate which is expected to lead to a doubled population in about thirty years. Few people believe that Third World economies will be able to grow fast enough to keep pace with such a rate of population growth. Already there are something like a billion people living in what Robert McNamara, former president of the World Bank, has aptly described as "absolute poverty."[6] The absolutely poor suffer from chronic malnutrition, a staggeringly high infant and child mortality rate, and a greatly reduced life expectancy for adults. Rapid population growth is likely to defeat all efforts to improve the quality of their lives.

One hopeful sign is that the annual world growth rate has declined from about 2 percent in the mid-1970s to about 1.7 percent at present. This has led some to conclude that the population problem has already been solved. Others deny that population growth is, in itself, a real problem. Optimists argue that the earth could adequately support many times its present human population. Whether or not that is true, it is clear that the world's present level of food production would be more than sufficient to feed its present population, if only the wealthy nations would stop wasting so much food (e.g., by feeding grain which is fit for human consumption to meat-producing animals), and if only we could create a more equitable worldwide system of food distribution.[7] It is also true that high population densities in the Third World are much less destructive of the global ecology than is the economic development of the wealthy nations, with their excessive consumption of resources and industrial contamination of air and water.[8]

Nevertheless, rapid population growth cannot continue indefinitely. The potential for expanded food production is finite. Even present levels of production may not be sustainable in the long run. The high productivity of, e.g., American farmers is based on an intensive exploitation of agricultural land, which is resulting in a steady loss of top soil and fertility. This loss can only temporarily be masked by the increased use of fertilizers. Most currently untilled land is untilled for a good reason, such as the lack of water. Large irrigation projects are expensive, environmentally destructive, and often lead to salinization (i.e., the buildup of salt in the soil) and eventual loss of fertility. Global changes in climate could easily reduce agricultural productivity. Thus, unless population growth is slowed further, there are apt to be chronic famines even more massive than those currently occurring in the sub-Saharan regions of Africa. Even a global redistribution of wealth can only postpone the day of reckoning, if rapid population growth continues. In the words of Isaac Asimov,

> Population growth at current rates will create a world without hope, gripped by starvation and desperation. It will be worse than a jungle because we have weapons immensely more destructive and vicious than teeth and claws.[9]

In the light of this situation, many have argued that sex selection will benefit humanity by producing lower birth rates. In *The Population Bomb*, Paul Ehrlich proposed the establishment of a Bureau of Population and Environment which would (among other things) encourage research on sex selection. In his view,

> if a simple method could be found to guarantee that first-born children were males, then population control problems in many areas would be somewhat eased. In our country and elsewhere, couples with only female children "keep trying" in hope of a son.[10]

Of course, parents who have only sons may also keep trying in the hope of a daughter. Studies have shown that American parents with either two sons or daughters are slightly more apt to say that they intend to have additional children than those who have one son and one daughter.[11] However, as Peter Singer and Deane Wells note,

> Sex selection would have an especially pronounced effect in countries like India, where many couples try to have at least two sons; sex selection would halve the number of children the average couple need to produce in order to have two sons.[12]

These authors emphasize the effect upon birth rates in the first generation: the average family would probably be smaller, because fewer children would be produced in the attempt to achieve a family of the desired sex composition. Others have pointed out that the birth rate in successive generations might be even more dramatically affected. In an article titled "Bat's Chance in Hell," John Postgate argues that sex selection can save the world from overpopulation. He observes that son-preference tends to be strongest in precisely those parts of the world which are most impoverished and which retain the highest birth rates. Thus, he argues, an inexpensive method of sex selection would greatly reduce the number of women in the Third World, thereby producing drastic reductions in the birth rate in the next generation.[13] Reasoning along the same lines, Clare Booth Luce says that,

> The determining factor in the growth of *all* animal populations is not the birth rate of *offspring*, but the rate of *female* offspring ... The world fertility rate today is 4.4 births per woman. Roughly half of these babies are females. Consider that if only one female baby were born per woman, even if the birth rate remained 4.4 children per woman, the world population growth rate would presently be reduced to zero. And if the world birth rate were only one female baby per *two* women, world population, instead of doubling, as it is now doing, every 34 years, would *un*double every 35 years.[14]

Postgate and Luce believe that sex selection is not merely *one* possible solution to the threat of overpopulation, but the only feasible solution. Its primary advantage, in their view, is that it would require no compulsion: "it would work with, not against, the cultural attitudes of parents, especially in the overpopulated countries."[15] Neither believes that the social effects of extremely high sex ratios would be particularly severe. Postgate surmises that women might be subjected to a form of purdah and deprived of the right to work or travel alone, while male homosexuality, polyandry, and mechanical and pictorial substitutes for heterosexual intercourse would become more common. However, he regards these changes as "matters of taste rather than serious concern." [16] Luce anticipates an increasingly intense competition among men for the possession of women. But she believes that women would benefit from having "a greater choice among inseminators," and "might be able, for the first time in history, to dictate their own terms for the improvement of their living conditions and status."[17] She says that,

> whatever the adverse effects on society of lowering the rate of female births, they would be infinitely more tolerable than the horrors and catastrophes mankind is *doomed* to endure if it fails much longer to slow down the birth rate of females.[18]

There are several questions which must be asked about this line of argument. First, would sex selection really reduce the birth rate? Second, is it sexist and/or racist to argue for the reduction of Third World birth rates through the implementation of son-preference? And third, are there other ways of encouraging birth rate reductions which are morally preferable to the way suggested by Postgate and Luce?

It is not obvious that sex selection would lead to lower birth rates, either in the first or in successive generations. Edward Pohlman cites a 1942 survey conducted by J.C. Flanagan, in which several hundred American military officers and their wives were asked whether they would have had more or fewer children had they been able to preselect sex. In this survey, 11 percent of the men and 17 percent of the women said that sex selection would have led them to have larger families, while less than 1 percent of either men or women said that it would have led them to have smaller families. The responses of husbands and wives indicated a mean increase of .13 and .20 children per family, respectively.[19] Pohlman speculates that this surprising result might be explained by the fact that many of these people had already completed their families, and might have been more willing to contemplate adding another son or daughter than not having had one of their already existent children. Most surveys of preferences with respect to *future* children suggest that sex selection would result in smaller, not larger families.[20] Nevertheless, the results of the Flanagan survey show that we do not really know how birth rates would be affected by sex selection, even in the first generation. It is conceivable that,

> in a male-exalting culture parents would seize on the ability to have many sons without the risk of daughters, and would have even larger families than otherwise. Or parents might decide to have the same total number of children as otherwise, but to have them be of the "preferred" sex.[21]

What about the prediction that the implementation of son-preference in many of the Third World nations would so reduce the number of females born that the birth rate of the next generation would be greatly reduced? This prediction assumes that Third World people and their governments would accept an enormous increase in sex ratios, without instituting counter-measures. But would they? Feminist protests against the selective abortion of female fetuses have already occurred in India and would be likely to spread and intensify were that practice, or even the preconceptive implementation of son-preference, to greatly reduce the relative number of women. Men might also rebel against a practice which deprived them or their sons of wives, and hence of legal offspring.

The desire to have children and grandchildren is often very strong, as is the fear of being "outbred" by other nations or ethnic groups. Were sex selection to seriously threaten the reproductive potential of Third World nations, it might come to be seen—perhaps with good reason—as a plot on the part of the richer nations to maintain their dominance. In that case, it might be rejected, in spite of its short-term advantages for parents. In the predominantly Catholic nations, it may be rejected in any case.

Where son-preference is strong and religion does not inhibit the use of new reproductive technologies, the absolute prohibition of sex selection might be ineffective. However, a variety of less severe measures might be employed to prevent extreme sex-ratio imbalances. Couples might be permitted to use sex selection for the production of sons only after they have already had one or more daughters. Special allowances or tax incentives might be provided for those rearing daughters; or those opting for all-male families might be subjected to tax penalties or other disincentives. Singer and Wells suggest still another possible strategy:

> If the method of sex selection were available only through registered medical practitioners, control could be kept by setting up waiting-lists for those who wanted a child of the sex that was being chosen too frequently. Couples who were more interested in having a child soon than in having one of the right sex would drop off the waiting-list, and an even sex ratio would be restored without frustrating the desires of those prepared to wait.[22]

Perhaps none of these strategies would be effective. Couples determined to avoid a firstborn daughter might resort to black market purveyors of sex-selection methods. Many might prefer tax penalties to the cost of rearing daughters. If conditions become more desperate, female infanticide and lethal neglect will probably become still more common if sex selection is not available. Thus, sex ratios might increase in spite of the best efforts of governments to prevent this. Yet the net effects of sex selection upon the birth rate might still be negligible.

A severe shortage of women could conceivably lead to the acceptance of polyandrous marriages. A woman might be supported by several men, and expected to bear children for each of them. Thus, she might have more children than would be feasible in a monogamous marriage. Fertility drugs might be used to increase the number of multiple births. Single men or infertile women might purchase children from professional surrogate mothers, who would produce babies on a production-line basis. Nations which wish to encourage population growth might pay couples for having more children, making childbearing a more profitable occupation than most others which are available to women. The age-old tactic of encouraging births by suppressing access to contraception and abortion would be used in many nations if birth rates fell below what national leaders considered desirable.

Thus, the long-term effects of sex selection upon birth rates are unpredictable, and the scenario painted by Postgate and Luce might prove to be highly unrealistic. Realism aside, it is morally objectionable for citizens of wealthy nations to suggest that Third World nations ought to fight overpopulation through massive reductions in the number of women.

Postgate and Luce leave themselves open to the charge of racism—or at least national chauvinism—through their failure to mention the responsibility of the richer nations to cooperate in the creation of a more just worldwide economic order. Through his casual dismissal of the loss of women's rights as a mere matter of taste, Postgate also courts the charge of sexism. Both Postgate and Luce make the dubious assumption that the basic cause of rapid population growth in the poorer nations is "the unlimited fertility of women."[23]

In one sense, this is obviously true; if women did not have so many babies then populations would not grow so rapidly. But this assumption begs the question of *why* women in poor nations tend to have so many children. Postgate attributes the fecundity of Third World women to "ignorance and shortsightedness,"[24] while Luce asserts that nature has "programmed" women to have many children.[25] These explanations are not only patronizing, but ignore the fact that the current population explosion is largely the result of the impact of western culture upon Third World societies.

Most societies in the past have had reasonably effective means of preserving a balance between human population and available resources. In *Sex and Destiny*, Germaine Greer describes some of the ways in which this balance has been upset through contact with western culture. Western influence has led to the decline of such traditional practices as postpartum abstinence, prolonged lactation, folk methods of contraception and abortion, and of course infanticide.[26] While it may be impractical (and, in the case of infanticide, immoral) to advocate a return to such traditional methods of population limitation, Greer's point underscores the absurdity of blaming too rapid population growth in the Third World upon the existence of too many women. Sex ratios are just as low or lower in the industrialized nations, yet in most of those nations population growth has slowed dramatically.

However poor they may be, women do not have babies merely because that is what nature has programmed them to do. They have babies either because they have no choice, or because they regard children as an asset. Whereas traditional methods of fertility control are easily disrupted by contact with western culture, it is virtually impossible for new methods to be successfully imposed from outside a culture, except through immoral forms of coercion. Thus, for people of the wealthier nations to debate the means by which Third World peoples ought to reduce their birth rates may be viewed as a form of hubris. No morally acceptable program for birth-rate reduction can succeed without the active support of the affected population.

That said, it must still be pointed out that there are more promising strategies for bringing about necessary birth-rate reductions than using sex selection to reduce the number of women. Free legal access to contraception and abortion is essential, and unavailable in much of the world. But it will often be insufficient to end excessive population growth, unless it is part of a much more extensive program of social change. The perceived need to have many children in order that some may survive to support the parents in old age can be reduced by such elementary reforms as (1) improved medical care, to reduce infant and child mortality; (2) improved social security for the elderly; and (3) the reduction of unjust economic differentials through property and income redistribution. Primary and secondary education, especially for women, helps to delay the age of marriage, to create more positive attitudes towards family planning, and to improve the efficiency with which birth-control methods are employed.[27] (Improved education for men is less often associated with lower birth rates, in part because men are less often forced to choose between having more children and earning an income.) Expanded work opportunities for women tend to reduce fertility by providing alternatives to child rearing as a full-time occupation. And every move towards more equal social, legal, and economic status for women helps to undermine the presumption that it is only sons who can improve the family's economic status, and thus reduces the incentive to continue childbearing until there are "enough" boys.

Social changes such as these are necessary for reasons of simple justice, and ought to be supported even by those who see no need to curb rapid population growth. Given a coordinated national effort, most of these changes are possible even for the poorest nations. China has demonstrated that literacy, medical services, and knowledge of child care and nutrition can be improved without huge expenditures, if each person who learns teaches others.

Unfortunately, China has failed to bring about genuine equality for women, and thus son-preference remains an obstacle to the implementation of the one-child policy. Some highly coercive means of enforcing that policy have been instituted. Westerners have been horrified by reports of Chinese women subjected to almost irresistible pressures to abort second- and even third-trimester pregnancies.[28] All methods of population control which are less than fully voluntary are costly in human terms, and involuntary abortion is impossible to justify under any circumstances. Were overt coercion essential, mandatory vasectomies for men who already have their quota of children would be less objectionable, in that the physical trauma is less severe, and sentient human life is not taken. However, such extreme measures are not necessary, at least in nations whose population crisis is less acute than China's. There are many less drastic ways in which governments can provide additional incentives for lower birth rates—such as imposing economic penalties on couples with more than one or two children, or rewarding parents who opt for small families. Such methods are not ideal, because they are apt to penalize the children as well as the parents of large families. There is little justification for resorting to such less-than-fully voluntary strategies for birth-rate reduction so long as the economic inequities within societies which encourage high birth rates among the poor remain.

Yet even such semicoercive strategies may be less detrimental in the long run than the seemingly noncoercive strategy advocated by Postgate and Luce. That strategy would do nothing to eliminate the social causes of too-rapid population growth. Rather than recognizing sexual inequality as one of the most important causes of excessive birth rates, they advocate birth rate reductions through the exploitation of son-preference. Such a strategy is bound to prove counterproductive in the long run. And in any case (to borrow Postgate's turn of phrase), it may not have a bat's chance in hell of being accepted by Third World people or their governments.

Greater Happiness for Parents

Perhaps the most obvious potential benefit of sex selection is that it will add to the happiness of parents, who will be able to control not just the number and spacing of their children, but also their sex. For many prospective parents, sex-preferences are extremely important, and their frustration a source of great grief. The satisfaction of such deeply felt desires is surely a point in favor of sex selection. Changes which make parents happier are likely to benefit children as well. But for the moment let us concentrate on the argument that sex selection will add to the happiness of parents.

There are reasons to doubt that sex selection will be as great a boon to parents as it at first appears. In the first place, couples may disagree about the ideal sexual composition of their families. The wife, for example, may want daughters for companionship and help with domestic work, while the husband wants sons as heirs or to help in *his* work. Regardless of whose wishes prevail, one partner may remain resentful and may take out that resentment on the other, or on the children. Thus, sex selection may open a new realm of potential marital conflict. Moreover, given the dominant position of the husband in most marriges, his wishes will often prevail, and many women will be deprived of daughters against their will.

In the second place, the attempt to preselect sex will sometimes fail. When this happens, one partner may suspect deception. In the absence of sex selection (it may be argued), they have only fate or biology to blame, but with it they may each blame the other. Women who fail to produce sons may be subject to increased abuse.

In the third place, parental expectations may be frustrated even if the child *is* of the preferred sex. If a son was wanted for material gain, which he fails to provide, parents may feel cheated. The power to select sex may encourage unrealistic expectations about the sex-selected child. Pressures to conform to traditional sex roles may be intensified if the child's sex was selected precisely because of certain features of those roles. When a sex-selected child rebels against such pressures, parents may be even more bitterly disappointed than had a child of the "wrong" sex been born.

These points show that sex selection will not be an unmixed blessing for all parents. But they do not prove that it will provide no net benefits for parents. As for the first problem, couples may also disagree about how many children to have or when to have them, but this does not show that access to effective contraceptives and to legal abortion is not on the whole beneficial. Women may also be coerced into accepting or refusing contraception or abortion; yet most would prefer to have such options available. After all, to be deprived of such options is also a form of coercion.

As for the second problem, the lack of effective means of prenatal sex selection in the past has not prevented spouses from blaming one another for their joint failure to produce children of the desired sex. Such blame is irrational since (so far as we know) sex cannot be determined simply by an act of will. Yet throughout history wives have been abused and husbands humiliated because of the birth of children of the "wrong" sex. There is little reason to suppose that "wrong" sex children who arrive because of the failure of sex-selection efforts will be any more unwelcome than those who arrive through reproductive roulette. Perhaps it will even be easier for parents to reconcile themselves to such "failures," given the improved chances for success next time.

What about parents whose sex-selected child does not conform to the script which they had in mind when they opted for a child of that sex? It is always something of a struggle for parents to accept their children for what they are, rather than what the parents hoped they would be. It is pure speculation to suppose that sex selection will make that struggle any more difficult than it has always been.

Thus, it seems likely that the power to select sex will be beneficial to parents more often than not. Getting what one wants is not a guarantee of happiness, but it is usually a good deal more conducive to happiness than *not* getting what one wants. However, the potential benefits of sex selection to parents are probably less significant than the potential benefits to children.

Every Child a Wanted Child

Many psychologists believe that individuals whose parents would have preferred a child of the opposite sex tend to suffer psychological harm. It has often been suggested that homosexuality or "sex-role confusion" can be caused by such parental disappointment, which may lead parents to treat a girl as if she were a boy, or vice versa.[29] Given that homosexuality is not inherently evil, and given that patriarchal sex roles have outlived any usefulness which they may once have had, perhaps we should discount these alleged psychological harms. Perhaps we should view anything which is conducive to sex-role "confusion" as a good thing. But parental rejection based on sex can blight a child's life even if it has no effect upon her "sexual identity." A girl who knows that her parents wanted a boy instead of her may believe that nothing she does will ever please them; and she may be right.

Some researchers have concluded that even a child who is not herself regarded as being of the wrong sex may suffer if other siblings are so regarded. If a girl is born first and the parents, hoping for a son, then have one or more additional daughters, all may suffer from the devaluation of femaleness. As Pohlman says,

> If a boy is born later, he may be unduly preferred in comparison to the other children. The others may be jealous of him. Even though his arrival rounds out the "one of each" desired, this may not correct all of the effects of the resentment felt toward the all-girl family before he arrived.[30]

Furthermore,

> Parents may be "forced" into having families larger than they otherwise would, and the capacity to provide an adequate emotional climate may be severely strained. This may have consequences for all family members, even those of the preferred sex who finally arrive.[31]

Hypotheses of this sort cannot readily be proven by quantitative research, but they are supported by clinical observation and everyday experience. A friend of mine who is the youngest of four daughters recalls that when she was a child people frequently expressed sympathy for her parents because they had so many daughters and no sons. At the age of thirty-six, she still feels pain at the way that her existence was constantly interpreted as a family misfortune.

This woman, however, was fortunate enough to have been born into a prosperous American family. The psychological harms she may have suffered because the sexual composition of her family was regarded as a misfortune pale by comparison to the suffering of unwanted daughters in poorer and more son-preferring societies. In much of the world, the difference between being a wanted daughter and being an unwanted daughter is often the difference between life and death. We will never know how many short and miserable lives will be avoided through sex selection, but the data on differential mortality rates for male and female children in northern India and many other parts of the world suggest that the number will be enormous. This is a benefit which does far more to counterbalance the possible ill effects of sex selection than the more speculative effects on birth rates or parental happiness. There is nothing speculative about the claim that millions of unwanted daughters have died or suffered abuse or neglect because of their sex. Fewer children will suffer this fate if better means of sex selection become available.

In practice, then, preconceptive sex selection or sex-selective early abortion will often function not as a form of gendercide, but as a practical alternative to gendercide. In an ideal world, all children would be equally loved and well cared for, regardless of their sex. Compared to that ideal alternative, avoiding the lethal neglect of female children by preventing the birth of unwanted girls leaves much to be desired. But compared to what is frequently the actual alternative it is a very great improvement.

Sex Selection as a Feminist Tactic

Those who believe that it is inherently sexist to preselect the sex of a child will be unlikely to view sex selection as a possible means of combatting sexism. But if sex selection is not inherently sexist then there is no contradiction in the suggestion that it may provide some women with new means of resistance to patriarchy.

In a strongly son-preferring society which might otherwise become predominantly male, the decision of some feminists to raise daughters rather than sons might be a necessary corrective. Some will refuse to raise sons because they are unwilling to contribute to the ruling sex class. Of course, the rearing of nonsexist sons is also a contribution to a more egalitarian future. But just as some women find it impossible to conduct sexual relationships with men without diluting their commitment to the women's movement, so some will judge that the attempt to rear nonsexist sons in a sexist society is a less than optimal investment of their time and energy. Some women will choose to have daughters as part of the project of establishing all-female families or communities. Others will choose daughters simply because they prefer the companionship of females. All of these choices may be forms of resistance to patriarchy. It is too early to predict that the choice of feminist women (and feminist men) to rear daughters in preference to sons will never become a significant aspect of that resistance. Already there are groups of women in England and America who are studying possible ways of conceiving daughters rather than sons.

Other feminists will use sex selection to have sons. Some will do so because they believe that they are particularly well qualified to raise nonsexist males. Some, finding themselves trapped in intensely misogynist societies, will be unwilling to bring female children into such a world.

There will also be many women who will "choose" to have sons because of direct or indirect coercion by their husbands or families. Medical practitioners who provide sex-selection services have a moral obligation to try to ascertain whether some of their women patients are being coerced into requesting a service which they would rather do without, and to withhold their services in such cases. Such a policy would be consistent with the practice of many feminist-run abortion clinics, which routinely interview their patients to be sure that the choice they are making is their own, and not made under duress. But it is a mistake to assume that feminist women would never preselect sons were it not for coercion. Choices which are conditioned by harsh social realities may still be authentic.

It may seem peculiar to suggest that some women will resist patriarchy by choosing to have daughters, while others may serve the same cause by choosing to have sons. But there is no contradiction, because there are differences in circumstances which make each choice rational in particular cases.

The use of sex selection may never by more than a minor part of the struggle against patriarchy. It is unlikely that enough women will opt for daughters in preference to sons to enable women to overthrow male domination through sheer force of numbers. In any case, sex selection can only be a means to an end. It is not a substitute for any of the substantive goals of feminism. To the extent that these goals are met, interest in selecting sex will decline. In the meantime, women will continue to resist male domination in innumerable ways, large and small; and these will sometimes include selecting the sex of their children.

Notes

1. Rayna Rapp, "XYLO: A True Story," *Test-Tube Women: What Future for Motherhood?*, edited by Rita Arditti, Renate Duelli Klein and Shelly Minden (London: Pandora Press, 1984), 321.

2. Laurence E. Karp, "The Prenatal Diagnosis of Genetic Disease," *Biomedical Ethics*, edited by Thomas A. Mappes and Jane S. Zembaty (New York: McGraw-Hill, 1981), 461.

3. See Helen B. Holmes, "Sex Preselection: Eugenics for Everyone?" forthcoming in *Biomedical Ethics Reviews, 1985*, edited by James Humber and Robert Almeder (Clifton, New Jersey: Humana Press, 1985).

4. Leon R. Kass, "Implications of Prenatal Diagnosis for the Human Right to Life," *Biomedical Ethics*, 467.

5. Holmes, op. cit.

6. Robert McNamara, *Summary Proceedings* of the 1976 Annual Meeting of the World Bank, 14.

7. See, for instance, Susan George, *How the Other Half Dies: The Real Reasons for World Hunger* (Totowa, New Jersey: Rowman & Allanheld, 1981); and Colin Tudge, *The Famine Business* (Middlesex, England: Penguin Books, 1977).

8. See Alfred Savoy, *Zero Growth?* translated by A. Maguire (Oxford: Basil Blackwell, 1975).

9. Isaac Asimov, quoted in *Time*, August 6, 1984, 12.

10. Paul Ehrlich, *The Population Bomb* (New York: Ballantine Books, 1971), 61.

11. See D.S. Freedman, R. Freedman, and P.K. Whelpton, "Size of Family and Preference for Children of Each Sex," *American Journal of Sociology* 66 (1960), 144. Data from the 1970 National Fertility Study support the same conclusion; see Charles F. Westoff and Ronald R. Rindfus, "Sex Preselection in the United States Some Implications," *Science* 184 (May 10, 1974), 633.

12. Peter Singer and Deane Wells, *The Reproduction Revolution: New Ways of Making Babies* (New York: Oxford University Press, 1984), 170.

13. John Postgate, "Bat's Chance in Hell," *New Scientist* 5 (April 1975), 12–15.

14. Clare Booth Luce, "Only Women Have Babies," *National Review*, July 7, 1978, 826.

15. Luce, 827.

16. Postgate, 15.

17. Luce, 827.

18. Ibid.

19. Flanagan, J.C., "A Study of Factors Determining Family Size in a Selected Professional Group," *Genetic Psychology Monographs* 25 (1942), 3–99; cited by Edward Pohlman, "Some Effects of Being Able to Control Sex of Offspring," *Eugenics Quarterly* 14, (4: December, 1967), 278.

20. Freeman, 141–46.

21. Pohlman, 278.

22. Singer, 171.

23. Luce, 824.

24. Postgate, 14.

25. Luce, 171.

26.   Germaine Greer, *Sex and Destiny: The Politics of Human Fertility* (London: Secker and Warburg, 1984).

27.   Susan Hill Cochrane's study of the effects of education upon fertility shows that in the poorest societies with the lowest level of female literacy, improved education often has the initial effect of *increasing* fertility. This is because education tends to increase the number of live births through improved health, better nutrition, and the abandonment of traditional patterns of lactation and postpartum abstinence. Eventually, however, female education generally reduces birth rates by decreasing desired family size, delaying the age of marriage, reducing infant and child mortality, improving communication between husband and wife, improving attitudes towards contraception, and increasing the perceived cost of children. The latter effect is particularly salient where there are nondomestic employment opportunities for women, as is more apt to be the case in urban areas. (Susan Hill Cochrane, *Fertility and Education: What Do We Really Know?*, published for the World Bank ([Baltimore: Johns Hopkins University Press, 1979])

28.   See, for instance, Steven Mosher, "The Truth About China," *The National Times*, March 2–8, 1984, 25–29.

29.   Pohlman, 277.

30.   Ibid.

31.   Ibid.

The Case for Freedom of Choice

Let me summarize the argument to this point. The nonconsequential-ist objections to preconceptive sex selection and early sex-selective abortion are invalid. The consequentialist objections cannot be as easily dismissed. However, we do not know whether the negative social consequences which many fear will actually occur, or how severe they are likely to be. Furthermore, there are some important probable benefits. The use of preconceptive sex selection or early abortion to avoid the birth of children with severe X-linked diseases, or children who will be abused or neglected because of their sex, would be a benefit of such magnitude as to outweigh many of the possible harms. Moderate sex-ratio increases may sometimes benefit heterosexual women by increasing their dyadic power, as Guttentag and Secord have argued. Moreover, some women will use sex selection as a way of furthering feminist goals. Thus, it is not clear that the possible negative effects are more significant than the possible benefits.

There is no formula which can enable us to weigh the possible negative effects of sex selection against the possible benefits, in order to determine which are morally more significant. It is certainly possible that the harms will outweigh the benefits. Some would argue that this possibility is enough to establish that sex selection is an immoral action and one which ought to be discouraged, if not by legal prohibition then at least by moral suasion.

But there are a number of reasons for resisting this conclusion. The first is that there is a general moral presumption in favor of freedom. This presumption cannot easily be overridden by a mere possibility of harmful effects—even rather significant ones. Second, the right of individual women and men to make their own decisions about the highly personal matter of reproduction is particularly vulnerable, and must be defended with special care. Finally, it is necessary to consider not just the possible negative effects of sex selection itself, but also those of the legal prohibition of sex selection. The social consequences of efforts to suppress the practice may well be worse than those of sex selection itself, were it to be freely tolerated. These points not only show that legal prohibition would be unwise, but also that the moral condemnation of sex selection is inappropriate.

The Presumption in Favor of Freedom

It is a truism that individual freedom of choice is a positive value, which ought to be preserved in the absence of powerful countervailing arguments. To prove that a particular action ought to be permitted, it is not necessary to demonstrate that toleration will produce net social benefits. Rather, the burden of proof is on those who would opt for legal prohibition. It is they who must show that prohibition would produce greater benefits than toleration. Even an action which is inherently immoral should not be legally prohibited unless there are good reasons to believe that prohibition will be beneficial.

Why is there such a presumption in favor of freedom? Why should we not make the opposite assumption, i.e, that unless there is persuasive evidence that the net effects of permitting a certain behavior will be beneficial, that behavior ought to be forbidden? The case for the presumption in favor of freedom has both a metaethical and a pragmatic component. At the metaethical level, this presumption is based upon respect for persons as autonomous agents. Mature, mentally competent persons are morally entitled to as much freedom as is consistent with the equal freedom, and the basic moral rights, of other persons. Furthermore, children and those who are not mentally competent are entitled to as much freedom as is consistent with their own well-being and the basic moral rights of others. There is room for endless debate about how much freedom is consistent with the rights of others. Yet the basic presumption in favor of freedom must be part of any satisfactory moral system.

The presumption in favor of freedom is sometimes criticized on the grounds that it ignores the social nature of persons, treating them as if they were isolated monads owing nothing to one another. Carried to extremes, or used as a rationale for the refusal to help those in need, the ideal of personal autonomy can indeed be antisocial. But the presumption in favor of freedom is entirely consistent with the essentially social and mutually interdependent nature of human existence. Our personal autonomy is in no way prior to our social existence, but rather grows out of it. If we were not social beings, and did not recognize obligations to one another, we would have no need for a concept of personal autonomy. We would simply do as we liked, without a thought about whether our behavior was a legitimate exercise of our own right to autonomy, or a violation of someone else's.

Although we are essentially social beings, we are also individuals with individual needs and desires. One difference between a human community and a community of social insects is that in a human community individuals come to recognize themselves as having interests which are not entirely submerged within the interests of the community. This is as true of socialist as of capitalist societies, though the line between individual and community interests may be differently drawn in the two cases. That is why, unlike ants, human beings require explicitly formulated systems of law and morality which not only articulate our duties to society and to other individuals, but also limit the extent to which society can require us to subordinate our individual interests to those of the social whole.

Freedom obviously cannot be an absolute value, because unless there are limits to what each individual is permitted to do, only the most powerful individuals will have any freedom at all. Nor can freedom be *the* primary social value; for unless basic human needs for food, shelter, health care, communal involvement, and so on are met, the abstract right to freedom becomes a mockery. Freedom is only one of many basic human needs, but it is not less basic than any other. The need for a measure of personal autonomy is as universal as the need for love, companionship, or physical security. However willing they might be to recommend such a life to others, few people would choose for themselves the life of a well-fed slave. And even well-fed slaves invariably spend a good deal of time and ingenuity finding covert ways of exercising their autonomy. Among human beings, mutually satisfactory social existence is difficult or impossible without respect for individual autonomy. Totalitarian systems of government are inimical to human well-being precisely because they unduly restrict the sphere of individual autonomy. That is why the

burden of proof is always upon those who would place further
limitations upon individual freedom.

Because freedom is a positive value, as well as a means for the
achievement of other positive values, it requires more than a significant *risk* of net social harm, should an action be permitted, to show
that that action ought to be prohibited. If the action can be shown to
be inherently wrong (e.g., because it violates basic human rights),
then it is not necessary to justify its moral prohibition by showing that
that prohibition will produce pragmatic benefits—although in most
cases it is reasonable to assume that it will. But even if an action is
inherently wrong, legal prohibition is inappropriate unless the probable harms resulting from legal toleration are demonstrably greater
than the probable costs of prohibition. Furthermore, the evidence for
these probable harms must be clear and persuasive; it must not be
based upon highly debatable religious, psychological or sociological
hypotheses. The hypothesis that the voluntary individual use of new
methods of preselecting the sex of children will be socially harmful is
as yet highly debatable. Thus, unless and until more substantive
evidence for that hypothesis emerges, it must be judged inadequate to
outweigh the presumption in favor of freedom.

So much for the metaethical argument for the presumption in favor
of freedom. One of the most important pragmatic arguments for
refusing to limit individual freedom without strong evidence of
significant benefits to be derived from that limitation is that the
enforcement of legal prohibitions is expensive. In every society, there
are limited funds and personnel available for law enforcement.
Resources devoted to the enforcement of prohibitions upon actions
which may not actually be harmful are inevitably diverted from the
enforcement of prohibitions upon actions which *are* harmful.

Another pragmatic argument against prohibiting actions which
are not demonstrably harmful is that people tend to ignore such
prohibitions. We will turn to that argument presently. But first, let's
look at a pragmatic argument which applies specifically to the
prohibition of sex selection.

The Defense of Reproductive Freedom

Even if some of the more pessimistic predictions about the consequences of sex selection begin to come true, there will still be reason to

resist prohibition. For there is great danger that the legal prohibition of sex selection would endanger other aspects of women's reproductive freedom. That freedom includes not only the right to choose abortion, but also the right to decide whether and when to become pregnant in the first place. It includes the right of women of all ages, married or single, lesbian or heterosexual, to equal access to both the older reproductive technologies, such as contraception and abortion, and the newer ones, such as artificial insemination and in vitro fertilization. It includes protection from involuntary sterilization, unnecessary hysterectomies and Caesarian sections, unsafe modes of contraception and similar medical abuses. And it includes a reasonable level of social support for child rearing. (This does not mean that all economic incentives designed to encourage small families are illegitimate, but only that it must not be unreasonably difficult for either single or married persons to rear their "fair share" of children—whether that be one, two, or more.)

None of these rights has as yet been fully secured. In many nations, all of these rights are under continual threat from right-wing factions. Under these conditions, any erosion of one element of reproductive freedom is likely to contribute to the loss of other essential reproductive rights. As Tabitha Powledge points out,

> To forbid women to use prenatal diagnostic techniques as a way of picking the sexes of their babies is to begin to delineate acceptable and unacceptable reasons to have an abortion. [This] argument against legal restriction of these technologies applies as much to pre-conception as it does to post-conception ones. [We must not] provide an opening wedge for legal regulation of reproduction in general.[1]

This is a slippery slope argument, and thus is superficially similar to some of the consequentialist objections to sex selection which we have rejected. According to one such argument, sex selection is dangerous because it is apt to lead to morally objectionable forms of genetic engineering. That argument is implausible, because people are quite capable of distinguishing between the different cases, and thus there is no reason to assume that acceptance of sex selection will lead to acceptance of less defensible practices. Is it any more plausible to argue that the prohibition of sex selection is dangerous, because any new limitation upon women's reproductive freedom is likely to lead to further and even more harmful limitations?

I think it is. The two cases are not parallel, for there is at present no

highly powerful interest group which is committed to the development and use of immoral forms of human genetic engineering.
Liberals and conservatives alike agree that genetic engineering must
not be used in ways which will be harmful to those persons whose
mental or physical properties are genetically altered. Thus, it is
implausible to argue that inherently harmless or beneficient reproductive technologies will inevitably give rise to harmful ones. There
are, on the other hand, extremely powerful religious and political
factions which are zealously opposed to most forms of reproductive
freedom for women.

American women won the right to legal access to contraceptives
only after a prolonged struggle against conservatives who predicted
dire consequences were women able to enjoy sex with less risk of
pregnancy.[2] Today, contraception and abortion are legally available
in the United States, at least to women who are no longer minors and
who can afford to pay for them. Yet there are thousands of individuals
and dozens of influential organizations which are determined to
eliminate women's legal right to choose abortion. These groups
employ the rhetoric of individual rights, arguing that abortion
violates the fetus's right to life. But their real priorities are revealed by
their opposition to the very things which are necessary to reduce the
need for abortion, such as better sex education in the schools and
easier access to contraceptives for teenagers. To conservative antifeminists, anything which increases women's control over their own
reproductive lives is anathema. Reproductive freedom is a threat to
the patriarchal family, which they confuse with family life itself. In
this political climate, the fear that antisex-selection legislation could
set a precedent for more severe limitations of women's reproductive
freedom is a realistic one.

These political realities also reveal the fallacy in the argument that
sex-selective abortion should be prohibited lest its occurrence undermine public support for the general right to choose abortion. Those
who support the goal of sexual equality, and who recognize that
reproductive freedom is essential to that goal, are not likely to turn
against freedom of choice simply because they disagree with some of
the ways in which it is exercised. Conversely, most of those who
oppose sexual equality are already opposed to the right to choose
abortion, and to other aspects of reproductive freedom as well. Surely
it is a wiser strategy to defend reproductive freedom in all its morally

legitimate aspects than to surrender part of that freedom in the hope of preserving the rest. We should argue against late abortion for the purpose of sex selection, but even in that case we should not support legal prohibition.

The Paradoxes of Unenforceable Prohibitions

There is one more argument against the prohibition of sex selection, which would be sufficient even if none of the preceding arguments were. It is that prohibition probably would not work, and would be likely to aggravate the very ills which it was designed to prevent.

Consider the class of so-called victimless crimes—e.g., gambling, prostitution, homosexual acts between consenting adults, and the production, sale or possession of pornography or of recreational drugs such as alcohol, tobacco, or marijuana. On the assumption that fetuses do not yet have a right to life, abortion also belongs on this list. These activities have several important features in common. First, they have been a common part of human life in many societies throughout recorded history. Second, they are all in one way or another connected with the pursuit of private satisfactions. (This is true even of abortion, which would rarely be necessary if people did not enjoy sex as more than a means of procreation.) Third, they are often condemned as immoral, not because they can be shown to violate the rights of other persons (for they cannot), but because they are presumed to be in some way harmful to society. As a result of this presumption, these activities have often been legally banned, though rarely with great success.

The argument against the prohibition of such victimless crimes is to some extent separable from the argument against regarding them as morally objectionable. Let us assume for the moment that all of these activities are in some way socially harmful. There is no doubt that all of these activities can be conducted in ways which are harmful. Gamblers may squander money needed for the support of their families; prostitution, like any other sexual activity, can spread disease; sadistic pornography may encourage violence against women or children; and excessive drug use may make the user a less productive member of society. But even if it could be shown that such ill effects are more significant than any pleasure or profit to be gained from these activities, it does not follow that banning them is a good idea. For the consequences of prohibition are often far worse than those of the activity itself, when it is legally tolerated and its commercial aspects regulated to minimize abuses.

The evidence for this conclusion is well known, and it would be tedious to go into it at length. A brief reminder of some of the ill effects of the prohibition of victimless crimes should suffice. These include:

(1) The reduction of resources available for the prosecution of much more serious crimes.

(2) The criminalization of persons who may do little or no harm to others or even to themselves.

(3) The subsequent undermining of respect for law in general.

(4) The exaggeration of whatever dangers are inherent in the activity itself; e.g., in the case of drug use, the danger of impurities or overdose.

(5) The provision of lucrative markets for crime syndicates, which sell products or services at inflated prices to persons who have no legal means of redress if cheated.

(6) The encouragement of police and judicial corruption through bribery and blackmail.

(7) The creation of systems of spies and informers to pry into private behavior.

(8) The encouragement of entrapment and the deliberate creation by the police of illegal acts.

(9) The creation of an illicit weapon (i.e., the easy fabrication of "evidence" of crime), which may be used against almost anyone.

(10) Encouragement of the prohibited activity itself, which may gain the lure of forbidden fruit, or come to be seen as a rebellion against the entire range of social injustices.

The factors which make it impossible to suppress most victimless crimes are the same factors which guarantee that legal prohibition will produce these counterproductive results. Because the act can be performed in secret, and because there is no victim to report it, it is difficult to detect its performance without deception or invasion of privacy, and easy to fabricate "evidence" where the act has not in fact occurred. The provision of the forbidden product or service is apt to be highly profitable, especially since prohibition inevitably produces artificially inflated prices. Thus there will almost always be people willing to provide it, however severe the penalty.

Because the forbidden actions are not in themselves overtly harmful, many people will inevitably continue to believe—with a good deal of reason and justice—that they have a right to perform those actions; and many will continue to do so regardless of the law. For the same reason, many law enforcement agents who would never take bribes to ignore genuinely criminal activity will do so in the case of victimless crimes. And many of them will commit these crimes themselves, thus becoming vulnerable to blackmail.

The prohibition of sex selection, like the prohibition of prostitution, would have especially deleterious effects upon women. Like female prostitution, the desire for sex selection is a symptom of the oppression of women in patriarchal society. Both are largely due to the absence of equal economic opportunities for women. (An analogous point may be made about the existence of young male prostitutes who cater to older male customers.) Prohibiting prostitution fails to alter the conditions which lead some people to sell sexual services and others to purchase them. It also magnifies the dangers and humiliations associated with the profession. Forced to operate on the wrong side of the law, prostitutes are less able to cooperate for their mutual protection, and are often forced to depend upon abusive pimps. (Of course, not all pimps are abusive, and not all prostitutes choose their profession because of economic need; yet these stereotypical notions are true in many instances.)

When prostitution is illegal, not only can prostitutes expect no protection from the law, but they must often bribe police to allow them to operate. Moreover, the stigma attached to their work victimizes all women. Any woman, however respectable, may be harassed, arrested, or denied service in public places if she dares to go out without a male escort. The relatively lenient treatment of men who purchase the services of prostitutes adds insult to injury. But punishing these men does little to improve the situation. Female prostitution is apt to exist wherever women are economically disadvantaged, and prohibiting it will rarely do more than multiply the associated problems.

The history of abortion provides an equally dramatic illustration of the paradoxes of unenforceable prohibitions. Legal access to abortion was strictly limited in most American jurisdictions prior to 1973, when The Supreme Court ruled unconstitutional the prohibition of first- or second-trimester abortion.[3] Yet millions of women nevertheless sought and obtained abortions. The difference is that illegal abortions were typically unsafe, frightening, humiliating, and excessively expensive. Too often, they were performed by unqualified persons under unsanitary conditions, with the result that each year thousands of women died of infection, hemorrhage, and other needless complications. Thousands more were left infertile or their health permanently damaged. Before surgical methods of abortion came into widespread use, women ingested an enormous variety of poisonous substances in order to end unwanted pregnancies; and many of these women also died or suffered permanent injury.[4] Legalization has almost eliminated deaths in the United States due to abortion, although the refusal of Medicaid funding for abortion has made it difficult for poor women to obtain safe abortions, and has resulted in some needless deaths.

Once a reliable method of sex selection becomes available, it is likely to be just about as difficult to suppress as any other victimless crime. Many prospective parents would not hesitate to defy legal prohibitions. Wealthier couples and single women would travel to those parts of the world where sex selection was legally available. But poorer women would be forced to seek treatment from black-market practitioners.

Illicit sex-selection treatments would inevitably be less competently administered and more dangerous, both for the woman and for the sex-selected child. Neither the law nor the medical profession would be able to supervise the conditions under which such treatments were provided, to prevent the use of unsafe or ineffective methods, or the coercion of women into accepting sex selection against their will. Police would collect bribes from some operators to look the other way, while setting up "sting" operations to entrap other providers and would-be parents of sex-selected children. Informers would be paid to pry into the private lives of parents who produced unisex families. The depressing spectacle of deceit, corruption and black-market profiteering which we have witnessed in connection with the prohibition of abortion, prostitution, homosexuality, pornography, alcohol, marijuana and "hard" drugs would be replayed in yet another arena.

If this scenario is implausible it is not because the prohibition of sex selection would be unlikely to produce such results, but rather because at present it does not appear likely that it will be prohibited, at least in the United States. As yet there is no strong movement for the prohibition of further research or marketing of new methods of sex selection. This is in part because relatively few people realize that sex-selective abortion is already a reality and that more effective preconceptive methods may soon be developed, and in part because many of those who are aware of these realities regard sex selection as wholly positive in its implications.

Sex selection has aroused little right-wing opposition, perhaps because—unlike the classic victimless crimes—it is not associated with the pursuit of what are often regarded as immoral pleasures. Nor does it, at first glance, appear to threaten the patriarchal family. Unlike surrogate motherhood or artificial insemination by donor, it does not introduce legal complexities about family relationships. Although it can be used to serve feminist ends, it is apt to be used more often in ways that pose no threat to male domination, and may even strengthen it. So it is not surprising that, thus far, the strongest opposition has come from feminists.

But this situation could change. If sex-selective abortion becomes increasingly common, antiabortion groups are likely to make a public issue of it. If the implementation of son-preference threatens to produce a shortage of women, or if many single women choose to have daughters, conservatives may come to see sex selection as a threat to the established social order. That is why it is important that feminists address the issue of prohibition now, before it becomes a heated public issue, and one which might be a source of division within the women's movement.

The Value of Moral Toleration

These are powerful arguments against the *legal* prohibition of sex selection. But are they relevant to the question of whether sex selection should be condemned as *morally* wrong? As Tabitha Powledge points out, moral exhortations may be effective even in such a personal matter as reproductive behavior. In her view, "We must say over and over again to friends and neighbors, in the pages of magazines and newspapers, on television and radio, that this technology, even if available, should simply not be used."[5]

There is no contradiction in the claim that although sex selection ought not to be legally prohibited, nevertheless it is morally wrong.

There are any number of morally objectionable actions which are and ought to be entirely legal. For instance, it is immoral to make cruel and gratuitous remarks about other people's appearance, but few would argue that it ought to be illegal. One reason why not all immoral actions ought to be legally prohibited is that some are too minor to be worth the cost of imposing legal sanctions. However, some actions which are and ought to be legally tolerated are more than trivially immoral. A cruel remark, for instance, may cause great pain. Another reason why the law ought not to prohibit all immoral actions is that some moral rights—such as the right to speak one's mind—comprehend actions which will in some instances be morally objectionable. Powledge's position is that sex selection should be condemned as universally wrong, but legally tolerated as an exercise of the right to reproductive freedom.

However, I would argue that the presumption in favor of freedom applies not only to legal but also to moral sanctions. This presumption should lead us to reject the claim that sex selection is inherently immoral. Respect for personal autonomy requires that moral condemnations be directed only against actions which are demonstrably harmful, and/or which clearly violate the rights of other persons. There is no evidence that all uses of sex selection are harmful. Violations of rights are not necessarily involved, except when people are coerced into accepting sex selection. Legal regulation of the practice and voluntary self-regulation by those who provide sex-selection services would probably be more effective in preventing such coercion than would moral condemnation of sex selection itself.

Of course, even voluntary sex selection will sometimes be morally objectionable, e.g., because it is done for sexist reasons or because the predictable consequences in the particular instance are on balance harmful. I have also argued that late-term abortion for the purpose of sex selection is morally objectionable, even though it is an action which women have the right to perform, and even though it is usually wrong to blame individual women for resorting to this means of sex selection. However, wholesale moral condemnation of all forms of sex selection is likely to be counterproductive. In a world in which overtly gendercidal crimes against women remain common, feminists cannot afford to waste moral capital by denouncing forms of behavior which merely *might* prove harmful to women.

Furthermore, when it comes to victimless actions which many

people are strongly motivated to perform, moral condemntion is very nearly as impotent as legal prohibition. Condemning drug use, prostitution or abortion as immoral does little to minimize their occurrence. At best, such moralistic pronouncements only lead to unnecessary guilt feelings. At worst, they undermine respect for morality as an objectively based system of principles and attitudes which is essential for cooperative social existence. People who are constantly told that actions which are not demonstrably harmful are nevertheless immoral are likely to conclude that morality is little more than a matter of subjective opinion, and to behave accordingly.

Those who believe that the world would be a better place if certain victimless "crimes" occurred less often should look for ways to alter the social conditions which lead to their occurrence, rather than condemning the individuals who engage in them. Sex selection will lose its attraction only when sexism is eliminated from our institutions and ideologies. It is a *symptom* of sexism in society, but not necessarily an *instance* of sexism. Condemning it as inherently immoral will only distract attention from more vital issues of justice and fairness. If and when more persuasive evidence of its harmfulness emerges, that will be time enough to reassess this moral stance.

Notes

1. Tabitha M. Powledge, "Unnatural Selection: On Choosing Children's Sex," in *The Custom-Made Child? Woman-Centered Perspectives*, edited by Helen B. Holmes, Betty B. Hoskins and Michael Gross (Clifton, New Jersey: The Humana Press, 1981), 197.

2. See, for instance, Linda Gordon, *Woman's Body, Woman's Right: A Social History of Birth Control in America* (New York: Penguin Books, 1974).

3. "Majority Opinion in *Roe v. Wade*," in *Contemporary Issues in Bioethics*, edited by Tom Beauchamp and Leroy Walters (Belmont, California: Dickenson Publishing Company, 1978), 243–46.

4. See James C. Mohr, *Abortion in America: The Origins and Evolution of National Policy, 1800-1900* (New York: Oxford University Press, 1978).

5. Powledge, 198.

Conclusion

Given the history and continued occurrence of anti-female gender-cide, the risks associated with the development of new means of sex selection must be taken very seriously. But these risks do not warrant either the legal prohibition or the categorical moral condemnation of these new means of sex selection. The only new form of sex selection which is inherently objectionable is late-term abortion. Late abortion performed for the purpose of sex selection is objectionable not only because it involves the taking of sentient human life for reasons which are based upon unjust sexual discrimination, but also because of the severe physical and mental trauma inflicted upon women by such abortions. Yet women have the moral right to choose abortion, and should not be blamed for consenting to sex-selective abortion. Objectionable as late-term sex-selective abortion is, it may sometimes be the lesser evil.

If effective and inexpensive means of sex selection become widely available, sex ratios will certainly be increased in many son-preferring societies. However, we cannot predict just how rapid or how extreme these increases will be, or how they will affect women's lives. This will depend upon the responses of individuals and governments, and we can only guess what these responses may be. Moreover, even if we could be certain that the results of sex selection will be on balance detrimental, there would still be reason to reject the conclusion that it should be legally prohibited. Prohibition might only produce a black market in sex-selection services, which would be immune to regulation either by legislation or by the official ethical standards of health-care practitioners. Such regulation may at least prevent the worst abuses of the new means of sex selection, such as the coercion of women into submitting to sex-selection treatments against their will.

These conclusions do not imply that feminists should accept the new methods of sex selection passively. There is much that can be done to ensure that sex selection will be used by and for women and not just against women.

More feminist women and men must become involved in the development and implementation of the entire range of new reproductive technologies. So long as these programs are run primarily by men with little feminist awareness, the techniques which are developed are apt to be made available only to married, heterosexual women. Furthermore, those techniques which have the greatest revolutionary potential may not be developed at all. Much research in sex-selection techniques, e.g., methods of sperm separation, has focused only on ways of increasing the odds of male conceptions. This is in part because the demand for sex-selected sons is greater than the demand for daughters. However, the fact that most of the researchers who are developing these techniques are nonfeminist men is surely also relevant. Little work has been done towards the development of techniques of cloning, parthenogenesis, or ovular merging, which would be particularly valuable for single and lesbian women. The prospect of such technologies is frightening to many people because they erroneously assume that such technologies would inevitably lead to gross abuses. But these lines of research have also been neglected because of the uncritical assumption that children ought to be produced and reared only by heterosexual couples. Such technologies could help to undermine the hegemony of the patriarchal family.

It will be difficult for enough women to become involved in the development of new reproductive technologies to make a real difference, unless fundamental changes are made in the medical profession and in government and other agencies which fund research. It is necessary for women of all ages and all ethnic groups to be involved in this research, and in the provision of the new reproductive technologies to prospective parents. They must also have the autonomy to set their own priorities. Reproductive services must be made available to all, without prohibitory fees, and without discrimination on the basis of marital status or sexual orientation. To accomplish this, it may be necessary for feminists to establish independently operated and funded research and clinical institutions. There are already feminist-operated clinics offering fertility control services, but there is a need for many more. There are enough feminist women and men with the necessary skills and resources to make this dream a reality, if its importance is recognized.

At the same time, we will have to put increased pressure on public and private funding agencies to support feminist researchers in the field of reproductive technology. All research programs in this field should have women actively involved in the setting of priorities. We must inform ourselves about the existing research programs and continually monitor them—from inside where possible, but from outside where necessary. We must protest discrimination against single and lesbian women in the provision of artificial insemination, in vitro fertilization, sex selection, and other reproductive services. We must insist that research on sex selection give equal priority to methods of selecting for females, and that those providing sex-selection services make both forms of selection available.

We must also do further research on the consequences of (certain uses of) the new methods of sex selection. We need to know more about how having older brothers affects girls in various cultures before we can conclude that the use of sex selection to produce first-born sons will be detrimental to women. If evidence that it *is* detrimental emerges, we will need to publicize this fact. We will have to try to persuade parents not to use sex selection in this way, and providers of sex-selection services to discourage their use for this purpose.

We will also need to monitor the consequences of the use of the new methods of sex selection throughout the world. Where sex ratios increase because of the implementation of son-preference, we must study the effects of these increases. Where these effects are detrimental, we will have to argue for self-regulatory practices by those providing sex-selection services, to reduce the impact on sex ratios. They might, for example, counsel parents not to have male firstborns or all-male families. Or, they might adopt Singer and Wells's suggestion of setting up waiting lists for couples requesting son-selection, in order to keep the number of sex-selected sons in line with the number of sex-selected daughters. If such self-regulation fails to prevent excessive sex-ratio increases, we may need to support government-created incentives for rearing daughters. We will also need to press for the inclusion of preconceptive sex selection and/or early sex-selective abortion in private and public health insurance programs, lest sex selection become a prerogative of the wealthy, and sex ratios increase disproportionately in the privileged classes.

But such strategies will probably not be enough to prevent dramatic sex-ratio increases in highly son-preferring societies should inexpensive methods of sex selection become available. Ultimately, the only way to avoid excessively high sex ratios will be to eliminate the cultural and economic bases of son-preference. This will require nothing less than the elimination of patriarchy itself. The repeal of sexually discriminatory laws is an essential first step. Laws requiring that wives and children bear their husband's or father's surname, and/or that property be inherited through the male line, have to be changed. Patriliny is often maintained by tradition rather than by law, yet most people assume that it is a legal requirement. Often, it simply does not occur to them that things could be done in any other way. Education and the influence of example may help to undermine such patriarchal traditions. Women must have the right to retain their own name after marriage and pass it to their children, should they wish to do so. They must have an equal right to inherit, own, transmit, and manage property.

Equal education for females is essential, and yet to be achieved in most of the world. Even in nations where sexual discrimination in education is theoretically banned, subtle and not-so-subtle pressures often operate to steer women away from scientific and technological studies. Affirmative action programs are necessary to counteract continuing sexist and racist discrimination in hiring and promotion, and to encourage women to seek careers in higher education and other predominantly male professions. We must continue to make the point that affirmative action does not mean reverse discrimination or the hiring of unqualified persons, but rather the correction of ongoing patterns of discrimination against women and minority people.

In every society, employment opportunities for women must be improved, and women's wages must be made commensurate with those available for men. The formula of equal pay for equal work is inadequate to correct salary discrimination against women, since employers need only assign men and women to different jobs (or to jobs with different titles) in order to circumvent equal pay laws. What is needed is a throughgoing reevaluation of pay scales, both to reduce excessive differentials between more and less prestigious forms of labor and—to the extent that such differentials are retained—to pay women in accordance with their actual levels of skill and responsibility. Part-time work needs to be available, to enable both women and men to combine paid labor with child rearing. Part-time work must be fairly paid, and part-time workers must have appropriate opportunities for job security and advancement. Child-care facilities must be provided at reasonable cost, to enable those with child-care responsibilities to take advantage of employment opportunities. Maternity and paternity leave must be available without loss of job security or seniority.

Such changes will not come easily. Much of the world is experiencing a conservative backlash against feminism, often spearheaded by fundamentalist religious leaders. We can either move forward toward the goal of human moral equality and harmonious coexistence with the rest of the natural world, or backward into a new dark era of patriarchal repression. Feminists, socialists, ecologists, gay-rights activists, peace and antinuclear activists, and those struggling against racial oppression and international imperialism must realize that their causes are mutually interdependent. It is difficult to move towards feminist or other progressive goals while poverty persists, and while scarce resources are poured into military armaments. So long as sons are a necessary form of social security, son-preference will remain strong. The unjust economic disparities between industrialized and Third World nations will have to be rectified if international peace is to be secured. Neither peace nor economic security will be possible if population growth and the destruction of the natural world continue at their present rate.

In short, we cannot hope to eliminate sexism and son-preference without surmounting a vast array of other problems. The implications of sex selection are problematic because patriarchy and other forms of injustice have not yet been overcome. The more that sexual and economic oppression persists and intensifies as populations grow and disparities between rich and poor nations and classes widen, the more likely it will be that sex selection will result in enormously high sex ratios in some Third World countries.

If very high sex ratios come about because of social conditions which virtually force parents to opt for sons, and if these high ratios intensify the oppression of women, then this process will have to be regarded as a form of gendercide. Yet sex selection will also be used as a means of resisting patriarchy, and as an alternative to some of the more blatant forms of gendercide, such as female infanticide and the neglect of female children. That is why we should not view it as an essentially negative development.

The fear of gendercidal uses of sex selection is in some ways analogous to the fear of genocidal uses of contraception or abortion. Both fears are sometimes fully justified. But those who argue that even the voluntary use of contraception and abortion is genocidal are much too eager to protect the right to have children at the expense of the right not to have them. Poverty and racial or sexual injustice may prevent women from having children which they would otherwise like to have. But in such cases it is these injustices which are genocidal, and not the voluntary use of birth control methods. Women have the right not to be coerced into the use of contraception or abortion, but they also have the right not to be forced to bear children against their will. There is no need to sacrifice the one right to the other. Reproductive freedom requires the simultaneous defense of both rights, and simultaneous opposition to both forms of coercion.

By the same token, it is neither necessary nor desirable to defend women's right not to be forced to use new methods of sex selection at the expense of their right to voluntarily choose to do so. Instead, we should seek to prevent overtly coercive abuses of the new technologies of sex selection, and continue to struggle against the injustices which perpetuate the preference for sons. Injustices such as the greater earning power of males may virtually force parents to prefer sons, and thus may be covertly gendercidal. But forcing parents to produce children who are unwanted because of their sex is no solution to the problems of sexual inequality. We must not accept the argument that women who opt for sex selection because of unjust patriarchal institutions are not making *real* choices. Often, they are making the best choice they can, under the circumstances in which they find themselves. As long as economic and political institutions are dominated by men, both overt and subtle forms of gendercide will continue to occur. But we must not seek to counteract gendercide through the erosion of reproductive freedom. It is up to those who recognize the pernicious effects of sexism, and its connection to all other forms of oppression, to work for the social changes which will enable sex selection to function neither as an indirect form of gendercide nor as a necessary alternative to more direct forms of gendercide.

Index

DATE DUE

MAY 1 1 2010			